Praise for *This Book I:*

"Not only is this book from times study of how past, present and future and even altered. *Back to the Future* and H.G. Wells's *The Time Machine* might not be mere fiction, after all!"

—Nick Redfern, author of *The Pyramids and the Pentagon*

"The question of 'What is Time?' has challenged serious thinkers throughout history, but the matter of Time Travel seems to be illusory and by its very nature prohibitive, not to mention being loaded with all sorts of internal self-contradictory enigmas. Can the past truly exist beyond our memories, and can the future exist beyond our dreams and the inevitable progress of linear time? Having come to know well Larry and Marie, these two stalwart explorers of the perplexing and the unknown set about dissecting the enigma of time travel with the same gusto with which they write all of their books. *This Book Is From the Future* is a thorough and fascinating work, dealing with some very weighty concepts and provocative theories in easy to understand layman's terms. The book is thoughtful, never extreme in its claims, but, at the same time, it never places restraints upon the reader's imagination. I heartily endorse this book as another of Flaxman's and Jones' exciting adventures of the mind."

—Brad Steiger, author *Mysteries of Time and Space*

"Get ready for your trip through time. *This Book Is From the Future* is the closest you'll get short of buckling up with Doc in his DeLorean. Marie D. Jones and Larry Flaxman have done it again, bringing their ability to make a subject approachable and entertaining while informing and educating at the same time. I'm buying this book for myself and my friends who are fascinated with the notion of time travel. H.G. Wells, eat your heart out!"

—Jim Harold, America's top paranormal podcaster,
Paranormal Podcast & Jim Harold's Campfire,
author of *Jim Harold's Campfire: True Ghost Stories*

"Is Time simply an uncontrollable progression that can be perceived and measured only by its usage? To the urban city dwellers caught up in the racing tumult of day-to-day busy-ness, time can move by pretty quickly, yet a pastural existence, free of the rushing elements of the clock moves by at a seemingly much slower pace. Are we hapless perceiver-experiencers, with Time being the quantity that spirals fleetingly through our hands? Or—as some purport—do we have the ability to become its practitioners? Despite the staggering paradoxes that could theoretically ensue, we have all dreamt of traveling through time in one direction or the other. Jones and Flaxman do what they have always done so well, this time taking us on a linear romp through Time Travel and deep into the Grid of the multiverse where they explore the stuffs of science fiction, to the quantum physics of scientific experimentation and metaphysical, mind-bending, worm-hole-sliding theory. The flux capacitor is warming up and DeLorean is running. Hop aboard for a tantalizing foray into Time."

—Scott Alan Roberts, author of *The Rise and Fall of the Nephilim*

THIS BOOK IS FROM THE FUTURE

A Journey Through Portals, Relativity, Worm Holes, and Other Adventures in Time Travel

By Marie D. Jones and Larry Flaxman

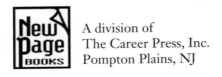

A division of
The Career Press, Inc.
Pompton Plains, NJ

THIS BOOK IS FROM THE FUTURE
EDITED BY JODI BRANDON
TYPESET BY DIANA GHAZZAWI
Cover design by Ian Shimkoviak/the BookDesigners
Photo on page 151 courtesy of Nick Redfern. Photo on page 172 courtesy of Sally Richards. Photos on page 187 courtesy of The Magic Castel and Starfire Tor. All other images courtesy of Wikimedia Commons.
Printed in the U.S.A.

To order this title, please call toll-free 1-800-CAREER-1 (NJ and Canada: 201-848-0310) to order using VISA or MasterCard, or for further information on books from Career Press.

The Career Press, Inc.
220 West Parkway, Unit 12
Pompton Plains, NJ 07444
www.careerpress.com
www.newpagebooks.com

Library of Congress Cataloging-in-Publication Data
Jones, Marie D.
    This book is from the future: a journey through portals, relativity, worm holes, and other adventures in time travel / by Marie D. Jones and Larry Flaxman.
        p. cm.
    Includes bibliographical references and index.
    ISBN 978-1-60163-150-3 (pbk.) – ISBN 978-1-60163-580-8 (ebook)
        1. Time. 2. Time travel. 3. Space and time. 4. Science—Philosophy. I. Flaxman, Larry. II. Title

Q175.32.T56J66 2012
530.11--dc23

                                                                2012009708

For Mary Essa and Max, future travelers in time....

ACKNOWLEDGMENTS

We would like to thank the entire staff of New Page Books for, once again, allowing us to put our work out in the world. We would also like to thank Warwick Associates for the awesome publicity. Most of all, we wish to thank our readers and our friends and colleagues for being so supportive and for pushing us to keep it real, even as we strive to set new bars to reach. Thank you for buying our books, listening to our radio shows, coming to our events, and offering your ideas, theories, and opinions. You make us better writers and researchers! And thanks to all of you who contributed material to this book—we deeply appreciate it!

Marie

I would like to thank my mom, Milly, for being my mom—and my number-one cheerleader. My sister, Angella, and brother, John, and my entire extended family. You guys keep me honest, laughing, and borderline insane, but I love it! I would also like to thank my dad, who is no longer here on earth, but hopefully traveling through time and space with eyes wide open. My family keeps me honest and anchored, and I love them all.

Thank you, Larry Flaxman, for being my dear friend, my business and writing partner, and my colleague, and for *always* inspiring me to be better and better not just as a writer and researcher, but also as a human being. You push me to think outside the box and always go that one level deeper, and I'd say after five books and five years, we make a damn good team, don't we? I am blessed to have you as a part of my destiny!

Thank you to *all* my friends. Writing is a very lonely profession, so having friends is so important and I am grateful for all of you, offline and online friends!

Most of all, thanks to Max, without whom this all would be quite pointless. I don't want to go back in time or forward in time because I am too busy cherishing every moment I have *now* with you, kiddo. You are my Multiverse!

Larry

First, and most importantly, I would like to thank my wonderful family, for without their support and constant encouragement, this book would have never come to fruition. To my mom, Sheila, my dad, Norman, my brother, Jon, and my wife, Emily: thank you from the bottom of my heart.

I would especially like to thank my daughter, Mary Essa, for providing me with the motivation and energy that I truly did not realize I had. Her boundless vigor, enthusiasm, and zest for life have absolutely helped to push me beyond my limits, and have eradicated the repetition and monotony of daily "normal" life. Mary Essa (aka "The Honey") is truly one in a million, and has proven to me that true love not only is possible, but it exceeds anything that I had ever before imagined. Joseph Addison said it best: "Certain is it that there is no kind of affection so purely angelic as of a father to a daughter. In love to our wives there is desire; to our sons, ambition; but to our daughters there is something which there are no words to express."

Many thanks to all that I've befriended throughout the years as a result of my books, research, speaking engagements, and radio and TV appearances. I can't even begin to express how grateful I am to those who have put their trust and belief in me! Thank you!

A special thank you to my partner and dear friend, Marie D. Jones. Words cannot express how incredibly honored I am to not only call you my business partner, but my friend. *Thank you!* It is obvious that our paths were fated to cross—and our work is the result of the beautiful and harmonious union of our strengths and beliefs. I can't wait to see what the next five years brings us!

CONTENTS

INTRODUCTION

Time travel impossible, say scientists.
—Discovery News, June 2011

Quantum time machine allows paradox-free time travel....
—Telegraph U.K., July 2010

Time travel theory avoids grandfather paradox.
—PhysOrg, July 2010

Could physicists make a time machine?
—Science Daily, August 2007

Wait a minute, Doc. Ah...are you telling me that you built a time machine...out of a DeLorean?
—Marty McFly in *Back to the Future* (1985)

On April 4, 2011, Handmade Mobile Entertainment Limited announced an iPhone app called "Flirtomatic," which allowed for, as they claimed, "location-based time travel flirting." With more than 4 million network members (and growing), Flirtomatic enabled its followers to make use of GPS and Google Maps to flirt with others by location. Though the service primarily focuses on real-time, instantaneous flirting, it also would allow you to do it utilizing a past location—or even one that you might visit at some point in the future! So, in other words, if you wanted to connect with someone you knew would be in a particular location two weeks from now, you could flirt forward. If you wanted to reconnect to a past love you once made out with in the high school baseball dugouts, you could set your high school as a past location in the hopes that your past love would find you and perhaps rekindle that old flame.

Meanwhile, another iPhone app allows people to time travel through specific locations, such as Bristol in the United Kingdom. The app is called "Time Traveler's Guide to Bristol" and was launched in June 2010. It features hundreds of archived images of Bristol as it was around 1910, and invites users to superimpose the archived images onto a 3D model of contemporary shots. This is a fascinating project and aims to show just how the city has changed...through time.

Okay, so it's not really time travel, but just the fact that technology is now jumping onto the time travel bandwagon certainly does say a lot about the public's ceaseless desire to transcend temporal limitations— not to mention how many people on Facebook, Twitter, and other social networking sites fell for the April Fool's joke courtesy of reporter Nick Hide, who on that fatefully fun day in 2010 wrote an article for the online Website Crave.net about a man arrested at the Large Hadron Collider in Switzerland for claiming to be a time traveler. According to the published story, one strangely dressed young man named Eloi Cole had traveled back in time to stop the Collider at CERN from bashing particles together to search for evidence of the Big Bang. His modus operandi was, according to Hide's piece, to stop supplies of Mountain Dew soda to the facility's vending machines. He was also reportedly rooting in trash bins at the Collider seeking fuel for his time machine power unit, which he described as similar to a kitchen blender.

Hopefully, most people got the joke right away (especially the reference to the blender—a la the "Mr. Fusion" utilized by Doc Brown in the movie *Back to the Future*), but from the speed by which the story spread that day and went "viral," it would appear that most people only read the headline before passing it on to friends!

The idea of being able to break out of the confines of the present and traverse time itself has long possessed, intrigued, and even obsessed the minds of scientists, writers, artists, and regular folks alike. Countless novels, books, movies, comics, songs, and television shows mirror our desire to once and for all rid ourselves of the chains of linear time. Time travel is very much a pop-culture icon.

Beyond the pop-culture aspect, from a theoretical science perspective, it's only a matter of time—literally—before we can travel back into the past, and forward into the future. Many researchers are already hypothesizing about what time travel might look like, and how we might achieve the ability to move beyond the limitations of the known laws of our physical universe. Perhaps the first step has already been taken, as scientists are finally coming to terms with the fact that these long-standing thought-to-be-unbreakable "laws" may actually not be as enduring as previously believed. Maybe we are already time traveling—but not in the way we think. Throughout the world, people from all walks of life have reported mysterious and bizarre experiences called "time slips" and "time shifts," where the temporal boundaries of reality are blurred beyond recognition. And other people report that they are, in fact, time travelers themselves, and guinea pigs for some top-secret government project that has succeeded in doing what some physicist still claim is impossible: placing a human being back into the past, or far into the future.

From *The Time Machine* by H.G. Wells, published in 1895, to the sci-fi best-sellers and blockbusters of today, time travel has enchanted us, drawn us in, and called to us, with the promise that one day the impossible will be proven possible. The imaginings will be made reality. Today's science fiction will become tomorrow's science fact.

Then we will all become potential time voyagers, traveling back and forth across the landscape of time itself.

Man...can go up against gravitation in a balloon, and why should he not hope that ultimately he may be able to stop or accelerate his drift along the Time-Dimension, or even turn about and travel the other way.
—H.G. Wells, *The Time Machine*

Fact or fiction: Are we able to time travel now? You might be shocked to learn that the answer to this question is a resounding *yes!* Believe it or not, scientists have already achieved successful time travel. Within the last five years, physicists have ostensibly become more interested in the concept and have even achieved time travel in controlled laboratory environments by sending a photon (light particle) into the future—well, a nanosecond of a nanosecond into the future. Although this is certainly extremely small scale, this, and others like it, proves that the concept is sound. All we have to do is find the mechanisms by which to make what occurs on a particulate level applicable on a grander scale—for human travel, a *much* grander scale. The behind-the-scenes science has become an increasingly important focus, and, throughout the past year, maverick scientists have experimented with ways to move beyond the confines of linear time. Meanwhile, in garages and basements and backyard labs, average Joes have tinkered with their own versions and visions of how to do just that.

Yet other physicists doing their own experiments insist that time travel will never happen.

Once confined to fantasy and science fiction, time travel is now simply an engineering problem.
—Michio Kaku, *Wired Magazine*, August 2003

But what if it could? Black holes, wormholes, time warps, time dilation, exceeding faster-than-light travel (FTL) speed, and alternate dimensions are all part of the current discussion of how time travel might become a reality. Stephen Hawking, one of the world's leading physicists, recently told the *UK Daily Mail*, "Time travel was once considered a scientific heresy. I used to avoid talking about it for fear of being labeled a crank…. But these days I'm not so cautious." Others, including physicist Michio Kaku, author of several books on everything from hyperspace to parallel worlds, are openly discussing the ways in which wormholes might be physically stabilized to allow for time travel. Others still focus their research on to the subatomic world of quantum physics for answers.

It seems that real science is finally coming out of the closet. In fact, as of this writing scientists at the Large Hadron Collider (LHC) at CERN in Geneva, Switzerland, are firing up the equipment once again to look for signs of something once thought of as sheer fringe sci-fi may become sheer fact: parallel Universes. The hundreds of scientists working at CERN reported to Reuters' Robert Evans for Yahoo! News, that they hope to use the head-on collisions of particles to prove that there are not only parallel universes, but also extra dimensions of space and time. Excuse us—extra dimensions of space and time? Imagine the implications. If this is true, imagine what doors this will open for time travel. We may not just be looking at traveling the landscape of time in our own universe, but in other universes as well, where the laws of physics just might allow for faster-than-light speed activity and beyond.

Hold on to your hats! This book will journey through the past, present, and future of serious time travel theories, research, and failures and successes, many of which may surprise readers who assumed the subject matter was simply the realm of fantasy and fiction. As scientific progress advances, new ideas and theories involving the possibility of manipulating the time-space continuum present themselves, and this book will examine them all, including cutting-edge theories and concepts such as the recent theory by astrophysicist Wun-Yi Shu of Taiwan that the Big Bang never happened. Shu's theory goes one step further and postulates that time and space are not independent of each other and can each be converted to the other. This new geometry of space and time is one of many cutting-edge attempts we will look at to first understand our universe, and then to control its laws to our advantage.

Other cutting-edge theories explore multiple Big Bangs, each with their own set laws of physics, many of which might allow for time travel. The Multiverse theory is not only catching fire in the world of sci-fi and fantasy, but in the world of hard science, where more and more physicists, cosmologists, astronomers, and astrophysicists are suggesting the presence of an infinite number of universes may not only explain cosmological mysteries, but quantum mysteries as well.

Today, we know that time travel need not be confined to myths, science fiction, Hollywood movies, or even speculation by theoretical physicists. Time travel is possible. For example, an object traveling at high speeds ages more slowly than a stationary object. This means that if you were to travel into outer space and return, moving close to light speed, you could travel thousands of years into the Earth's future.
—Clifford Pickover, *Time: A Traveler's Guide*

Ah, but it requires machines and devices, and we will explore the many attempts at creating time machines and devices that promised to advance time travel research forward, yet often failed to do so. Often, but not always—for some of the machines have led to newer, more workable theories that just might, in our lifetimes, find us sending far more than a photon into the future, or the past.

Notwithstanding the significant technical issues, the concept of time travel isn't completely casual and carefree. Of course, like everything else in life, there are paradoxes and inconsistencies to be addressed. Can you travel back in time, shoot your own grandfather, and still be born to travel back in time? Can you travel back before the inception of the time machine? These and other critical challenges to time travel will be detailed, including the potential for changing a fixed past, which has roots in quantum mechanics. We will even present a fictional scenario of an important event in past history, and propose how, if a time traveler were to go back and *change just one small thing*, that event might have played out instead.

This book will also delve into the many ways that we might already be traveling through time, as in chronesthesia, or mental time travel, dreams, déjà vu, missing time, time slips, and other paranormal experiences, and how perhaps time travelers have already contacted us in the form of ghosts, UFOs, and aliens, including the ongoing revelations from those in the know that the Roswell UFO crash cover-up may have involved human time travelers from our future. We will look at the many intriguing links to time travel and ancient religions and myths, and how time travelers might have appeared on earth before—or are here now. Could there be places on earth where people enter into time riffs and time storms, and actually go back or forward in time? Some researchers claim this has happened to hundreds, if not thousands, of people.

We may, it seems, even be able to be altered and affected by experiences that haven't even happened yet, as new research by professor of psychology Daryl Bem of Cornell University suggests. His work, recently published in the *Journal of Personality and Social Psychology*, shows that time is leaking and the future is slipping into the present. His studies, which we will document in this book, involved recall tests of specific, randomly chosen words that were scanned and typed by students as a computer selected them. The results showed that students were better able to recall words they had scanned and retyped *after the test* and, in his words, "that practicing a set of words after the recall test does, in fact, reach back in time to facilitate the recall of those words."

The idea is, then, that we can do something tonight that has already affected us today. Time is bleeding. The lines between past, present, and future are beginning to blur.

Spooky!

If we could travel into the past, it's mind-boggling what would be possible. For one thing, history would become an experimental science, which it certainly isn't today. The possible insights into our own past and nature and origins would be dazzling. For another, we would be facing the deep paradoxes of interfering with the scheme of causality that has led to our own time and ourselves. I have no idea whether it's possible, but it's certainly worth exploring.
—Carl Sagan, NOVA interview, October 12, 1999

This book will also look at the even more paranormal aspects of time anomalies, such as slips, warps, and shifts, and how these experiences might change our perception of time itself.

Among additional topics to be explored are:

1. Time and what we know about it. Is time fixed?
2. Why some physicists insist time travel is impossible.
3. Why wormholes, parallel universes, and extra dimensions might allow for space and time travel.
4. How people may already be traveling through time in their dreams, in paranormal experiences, and in altered states of consciousness.
5. The use of meta-materials to create a "space-time" cloak and how it might manipulate electromagnetic fields and create distortions of time.
6. Why forward time travel might be more probable than moving backward in time.
7. The roles of the Multiverse in making time travel a reality.
8. How some people insist they are time travel guinea pigs for a secret DARPA time machine experiment.
9. How time travel concepts relate to ghosts, aliens, and other conspiracy theories.

10. Time shifts and co-existing time lines. Are we alive in two different temporal dimensions?

11. Why the human brain is structured to perceive linear time and how that might be changing.

Our goal is to provide an exciting glimpse into the past, present, and future of time travel, without getting too deep into the physics and math. This is a book for the layperson, so we are not going to go all "deep physics" on you, and we happily provide enough information to tantalize you and hope you will make use of the resources in the Bibliography to further delve into the "deep physics" on your own. There are numerous amazing books, journals, and articles on the subject of time travel available through a variety of sources that can expand your knowledge and understanding of the hard science behind the theories. Please, make use of them.

This Book Is From the Future will be a journey itself, exploring the entire landscape of time, time anomalies, and time travel evolution: where we started. Where we now stand. What dreams may come. And why time travel is so much a part of the collective human psyche and the drive toward knowledge of the world we live in, and other worlds that maybe—just maybe—are waiting for us to discover, out there somewhere...in time.

Time is of your own making;
its clock ticks in your head.
The moment you stop thought
Time too stops dead.

—Angelus Silesius, sixth-century philosopher

1

WHAT TIME *IS* IT?

Clocks slay time...time is dead as long as it is being clicked off by little wheels; only when the clock stops does time come to life.
—William Faulkner

Time is but the stream I go a-fishing in.
—Henry David Thoreau

It is impossible to write a book about time travel without first addressing time itself. What do we really know about this crazy little thing called time? We know that time is an ingrained and significant part of our existence. Most of us live our lives according to the ticking of a clock, the number of days on a calendar, and the order of events that pepper our lives. We know that we had a past and that we are living in the present. We hope beyond hope that we will be around for the future. We want more time. In fact, most people will tell you that the two things they desire most in life are love and money. But when all is said and done, the reality is, we want more time to pursue love and money, and time to enjoy the fruits of our labors. Beyond this scant bit of knowledge, how much do we truly know about the mechanics of time?

Measuring Time

What exactly is time, and how do we measure it? Taken to the extreme, time to us is really nothing but a way of measuring what we are doing. It's a record-keeper of the events of our lives, all of which conspire to unendingly move forward in a linear fashion. This measurement of events, or intervals, as the Greeks suggested, can be broken down into smaller and smaller units, such as years, months, weeks, days, hours, minutes, seconds, and milliseconds. The measurement of time can also be built up into bigger and bigger units, such as millennia, eras, ages, and epochs. Technological progression has allowed us to design means to more accurately measure the passage of time. The invention of recording devices such as sundials and clocks are simply humanity's attempt to not only keep time, but also possibly to gain control of it. If we could tell how much time was passing, we would then be able to best determine how to use it—or not use it.

Units of Time

Unit	Size	Notes
yoctosecond	10–24 s	
zeptosecond	10–21 s	
attosecond	10–18 s	shortest time now measurable
femtosecond	10–15 s	pulse time on fastest lasers
picosecond	10–12 s	
nanosecond	10–9 s	time for molecules to fluoresce
microsecond	10–6 s	
millisecond	0.001 s	
second	1 s	**SI base unit**
minute	60 seconds	
hour	60 minutes	
day	24 hours	
week	7 days	also called *sennight*
fortnight	14 days	2 weeks
lunar month	27.2–29.5 days	Various definitions of *lunar month* exist.
month	28–31 days	
quarter	3 months	

year	12 months	
common year	365 days	52 weeks + 1 day
leap year	366 days	52 weeks + 2 days
tropical year	365.24219 days	Average
Gregorian year	365.2425 days	Average
Olympiad	4 year cycle	
lustrum	5 years	also called *pentad*
decade	10 years	
Indiction	15 year cycle	
generation	17–35 years	approximate
jubilee (Biblical)	50 years	
century	100 years	
millennium	1,000 years	
exasecond	10^{18} s	roughly 32 billion years, more than twice the age of the universe on current estimates
cosmological decade	varies	10 times the length of the previous cosmological decade, with CÐ 1 beginning either 10 seconds or 10 years after the Big Bang, depending on the definition

Courtesy of Wikimedia.

In our naive arrogance, we often believe that we have achieved ultimate mastery regarding the knowledge of our reality and very existence. The truth is that we know very little. Even if we think we know what space is, how does that correlate to our understanding of time? And if the two are intertwined (and many scientists believe they are), can changes to one affect the other? This creates a philosophical and scientific paradigm of sorts. If space always existed, then time must have as well—or did matter and time not exist until the moment of the Big Bang?

Maybe time is just change. Before the Big Bang, there was nothing. Afterward, there was expanding change as space and time came into existence. Yet does this mean that if the universe one day collapses back into the nothingness from which it came, space, but also time, would cease to be?

Maybe time is motion. In a fundamental sense, time measures motion, and motion is itself a kind of change. A change in the motion of an object implies that time has passed, the time from the moment the object was at one place, and then another, or the time that the object was one thing, and then became another.

But to keep it simple, time is how we give order to the sequence of events that make up our lives, and the duration of each event, including the amount of time between them. As science fiction writer Ray Cummings wrote way back in 1922, long before others were credited with his brilliant insight, "Time...is what keeps everything from happening at once!" We all buy into this concept, as in the Newtonian notion of "absolute time," in which time is considered universal, "with a unique, universally agreed upon notion of simultaneity of events and a unique, universally agreed upon time interval between any two events." That is how Kip S. Thorne described it in his book *Black Holes and Time Warps: Einstein's Outrageous Legacy*. Makes sense to all of us.

Is Time Real?

The problem is, time may not even be real at all. There was a time when science labeled time as fixed, real, and an inherent part of the cosmic structure. Time was considered to be a fixed tangible that was neither flexible nor changeable. Philosophy asked if there could be more than one kind of time—the real firm kind of time described by science and a fundamental property of the universe, and a time that related directly to light speed; and the subjective perception and experience of time. In the former, there is no implication of human alteration of time. We have no power and no control. The latter implied that time only goes as slow or as fast as we subjectively experience it, something we discuss in more detail in Chapter 8. The former was Newtonian time—the stuff of realists. The latter was the stuff of philosophers such as Immanuel Kant, who suggested that time was neither an event nor a thing. Kant believed that time could not be measured, and therefore could not be traveled. Are you confused yet? Seriously? You Kant be. (Sorry—couldn't resist.)

Philosophers struggled to deal with time long before scientists came along and forced them to rethink everything based upon new discoveries and understandings of the way our world works. Until that point, time was all about looking at the past as what happened and is now gone, the future as what will happen and has not yet arrived, and the present, which is the gift of the moment. The now. And the now was really all we ever had, was it not? The arrow of time shot forward, but the only true experience of time we ever had in a conscious sense was the present moment one. We love the way physicist Fred Alan Wolf describes this in his book *Time Loops and Space Twists* when he says: "We think of the past as having slipped out of existence, whereas the future is even more shadowy, its details still unformed. In this simple picture, the 'now' of our conscious awareness glides steadily onward, transforming events that were once in the unformed future into the concrete but fleeting reality of the present, and thence relegating them to the fixed past."

We love time. We hate time. And for the same reasons: because it seems out of our control, even as it gives order to our lives.

■■■■■ ■

Time for a Song

We spend so much time thinking about time, we even sing about time all the time! Remember some of these popular songs about the temporal dimension?

"Time in a Bottle"—Jim Croce

"Time After Time"—Cyndi Lauper

"Does Anybody Really Know What Time it Is?"—Chicago

"Time Is on My Side"—The Rolling Stones

"Time Won't Let Me"—The Outsiders

"Time Passages"—Al Stewart

"As Time Goes By"—Jimmy Durante

"Too Much Time on My Hands"—Styx

"It's the Most Wonderful Time of the Year"—Andy Williams

"Time Has Come Today"—The Chambers Bros.

"No Time"—Guess Who

"The Times of Your Life"—Paul Anka

"Time Warp"—*Rocky Horror Picture Show*

"Funny How Time Slips Away"—Willie Nelson

"Time of Your Life"—Green Day

"For the Good Times"—Ray Price

"More Today Than Yesterday"—Spiral Staircase

"Time"—Pink Floyd

"Crying Time"—Ray Charles

"Time of the Season"—The Zombies

"Second Time Around"—Frank Sinatra

"Funny How Time Flies"—Janet Jackson

"Remember the Time"—Michael Jackson

"Anytime"—Journey

"It's Too Late"—Carole King

"It Only Takes a Minute"—Tavares

"Seasons of Love/525,600 Minutes"—*Rent*

"The Age of Aquarius"—*Hair*

"Day by Day"—*Godspell*

"Time Is Tight"—Booker T and the MGs

"Tomorrow"—*Annie*

"The Way We Were"—Barbra Streisand

"It's Just a Matter of Time"—Brook Benton

"This Is the Time"—Billy Joel

"Yesterday"—The Beatles

"Turn! Turn! Turn!"—The Byrds

"Always and Forever"—Heatwave

"At Last"—Etta James

Meanwhile, religion stated that there might be different "times" for different cultures. The speedy life of busy New Yorkers might be measured in time far differently from the dreamtime of the Aborigines. Time might be linked to belief and action. It might be part of the Maya of illusion in the Buddhist tradition. Heck, even good ol' Einstein himself stated, "The distinction between past, present and future is only an illusion, even if a stubborn one." Thus, the timelessness often described by those who attain deep spiritual or meditative states may be nothing more than the lifting of this veil of illusion. Does time shape our concept of reality? Or does our reality shape our concept of time?

The Arrow of Time

Time is analogous to an arrow. Consider the idea of the arrow flying in one direction only. However, it is made up of three elements: past, present, and future. Once the arrow is set free, its trajectory cannot be changed, or can it? Something can step in front of the arrow (and hope it's not sharp!), suggesting that the future might be more malleable than we think (and possible in time travel terms). But the past trajectory cannot be changed, because it happened and is over and done with. The arrow, then, is really always only in the present because, as it moves, the past falls away and the future is about to become. The now keeps on going, as many metaphysicians tell us, and the present is really all there is. Using this analogy, it would appear almost as if time consists of an eternity of present moments. Nothing happens anytime but in the now.

In his book *Time Loops and Space Twists* physicist Fred Alan Wolf writes that "We usually put the facts of our lives into a tabulated form we call temporal order, tending to compare the instant moments of our lives with moments to come or those have gone by." He goes on to explain how we use space to help us do this tabulation, as when we say "I will meet you here, then." Then we can look around and see if where we are is the same or different from where we were or where we hope to be. And often we link space to time, as it should be, by saying things like "Meet me in an hour on the corner by the deli."

When Einstein came along, his theories of relativity and general relativity as well as his work with light and speed of light limitations forever changed our perception of and, ultimately, the overall face of time. Relativity states that two simultaneously occurring events, when observed from one point of view will appear to occur at different moments when viewed from another point of view, provided the second observer is moving relative to the first. Wolf explains that because of relativity, even the concept of a now is truly not an absolute, because "one person's now is simultaneously another's future and past, provided the two are simply moving relative to each other. Throw in the invariant speed of light, and things get even more interesting. Einstein postulated that the closer an object comes to traveling at the speed of light (186,000 miles per second), the more time would appear to slow down from the perspective of someone who was not moving in relation to the object. This slowing of time due to motion is called time dilation.

Time, motion, change, lightspeed, relativity. We'll revisit Einstein's views on time, and time travel, a bit more in depth later. First, let's explore a little bit more about time itself.

Is time a dimension? We know that we exist in a three-dimensional physicality. We have three spatial dimensions of height, width, and depth. So if time also measures something, like motion and change, does it by default become a dimension of its own? This fourth dimension is temporal, and yet directly interlocked with the spatial. Just imagine standing on a New York street telling a friend to meet you for pizza in an hour. You would tell your friend where you are in a spatial sense, but also what time to be there in a temporal sense. And if he or she got the spatial and temporal together according to your directions, you would enjoy a great Brooklyn-style pizza and an ice-cold beer with a friend.

An Ancient Look at Time

In my studies of ancient cultures, I was surprised to learn that many of them view time in ways that seem quite different from the ways I had been accustomed to conceptualizing it. For example, in modern Western cultures we typically think of time as a kind of linear progression, based on a future that lies *ahead of us* and a past that slowly trails out *behind us*. But in societies such as the modern-day African tribe from Mali called the *Dogon*, or the *Na-Khi* of Tibet and China, the instinctive conception is that the future lies *behind us* and the past ahead of us. Though this mindset may seem counter-intuitive at first, it makes better sense if we consider it in the proper way. For example, try to envision your ancestors and relatives as a long line of passengers riding together on a bus. Envision the oldest of your ancestors— your great- great-grandparents—getting off the bus first, followed by your great-grandparents, then your grandparents, then your parents, and, finally, you and your siblings, followed by your eventual offspring. Thought of in this way, it becomes much easier to imagine your ancestors (the past) as the ones who stand *before you* and your descendants (the future) as the ones who are destined to *follow you.*

Consistent with our modern notion of linear time, it was common among ancient societies in Europe, native tribes in the Americas, and other societies around the world to associate time with the image of a river. It is easy to see how the movement of time at its steady pace in only one direction might be compared to the flow of water from the headwaters of a river to its mouth. Likewise, time has been alternately associated with the concept of an arrow—an object that also moves in a straight line and at a steady rate, but again in only one defined direction.

Many cultures of the Eastern traditions conceptualized time not as a line, but as a circle. One of the earliest depictions of time—dating from the Neolithic era in China (around 6000 BC)—was as a serpent or a dragon, described in myth as eating its own tail. Similar concepts existed among other cultures such as the Egyptians and Hebrews and even persisted thousands of years later in Greek mythology, where the serpent was given the name *Uroboros,* derived from Greek words meaning "tail eating."

It is consistent with this worldview that we find concepts of time expressed in many of the earliest ancient hieroglyphic languages— including those of ancient Egypt and ancient China and Tibet—in words that center on the figure of a circle with a dot in the center. In some ancient religions, such as the *Vedic* tradition of India, the continuum of time is also measured in grand cycles that would be comparable to, but on another level of order greater than, the familiar *centuries, years, months,* and *days* of our everyday experience. These very large recurring periods (roughly 26,000 years) of the Vedic tradition are referred to as the *Great Year* and are thought to relate to a very slow apparent rotation of the constellations that astronomers call *precession.* The Vedic teaching is that humanity proceeds through long epochs called *ages.* During the very long cycle of the Great Year, humanity is thought to go through periods in which the capabilities of individuals increase and then diminish, similar to the normal daily periods of wakefulness, rest, and sleep that might occur in a normal day, and comparable to the familiar agricultural cycles of growth and dormancy that happen every year. These long stages of humanity are said by some to be governed by the conceptual "seasons" of this lengthy Great Year rotation.

Ancient European societies such as the Celtic culture are thought to have associated the symbol of the spiral with concepts of time. Accordingly, the Celtic year was described as "spiraling into autumn." The Tabwa tribe in Africa also conceived of time as a spiral and associated it with the shape of a shell that moved clockwise when viewed from the outside and counter-clockwise when viewed from the inside. It is interesting to note that that some modern-day geologists also find it useful to depict long periods of geologic time as a spiral.

Figure 1-1: The geological time spiral.

For many ancient cultures, such as the Vedic, Buddhist, and Dogon, traditional concepts of space and time were considered to be an illusion—a mere "reflection" or "image" (the Dogon say a "correct image") of a more fundamental reality that resides on another, more fundamental plane of existence. In fact, certain events that transpired during the creation of the universe and of matter were considered to be *eternal*, in that they effectively existed "outside of" the bounds of time, or were said to have come about "before" the creation of time. If our sincere belief is in time as an illusion, then it would seem that we might be able to train ourselves to see beyond that illusion—just as one can learn to "look past" an optical illusion and see the *vase* instead of the *faces*—and effectively step "outside" of the constraints of time. The Vedics actually have a concept that expresses this very notion called *kala-vancana*, or literally "time-cheating."

—Laird Scranton, author of *The Science of the Dogon* and other books on ancient cosmology and language

Do we somehow cheat time when we use Daylight Savings and gain an hour, or fall back and lose one? Is the tiny island of Samoa cheating time when in May 2011 it leapt 24 hours into the future by aligning itself with the weekday dates acknowledged in Australia, New Zealand, and Eastern Asia, thereby foregoing a 119-year tradition of aligning instead with U.S. traders in California? It was good business, after all, as its interests lay more with the Asia-Pacific region and not with California.

Can we decide that when we turn a certain age, to instead celebrate a birthday from the past over again? Always be 30 or 40? The same amount of time will pass to be sure, but if we convince others and ourselves enough, they might forget we were also 30 last year. Or 40....

Time's Forward Motion

In 1927, British astronomer Arthur Eddington coined the term *time's arrow*, to describe the immutable fact that time seemed to only be able to move forward, like an arrow shot into the air. Whether time was going fast or slow, it was still going forward. And unlike the three spatial dimensions, which can go up and down and side to side, the one dimension of time could only go forward. In his book *The Fabric of the Cosmos*, physicist Brian Greene, a professor at Columbia University, explained it by saying: "In space, we can move any which way we want, at will. Yet there's this other dimension to the universe that has an ironclad lock on us. It always drags you in one direction." Greene has been involved in research into time, space, quantum physics, superstring theory, and everything else of cosmic importance, as have many brilliant minds we will examine in this book.

In February 2010, Sean Carroll, a theoretical physicist at Caltech, spoke at the annual meeting of the American Association for the Advancement of Science on the topic of the arrow of time. Carroll suggested the concept of time as an arrow could be traced back to the ideas of Austrian physicist Ludwig Boltzmann. You might remember Boltzmann from Physics 101 as the scientist who figured out the law of entropy, which basically measures how chaotic or disorderly things are and how the chaos and disorder increase with time, becoming more disorderly. This is the Second Law of Thermodynamics. Carroll presented his research to the meeting attendees and also was interviewed at length for the February 2010 edition of *Wired* magazine. In the interview, he told reporter Erin Elba that he still felt something was missing from the relationship between entropy and time's arrow.

It all goes back, of course, to the Big Bang and why it had the properties it had—why the conditions of the creation of our universe were just as they were—theorizing that there might have been something before the Big Bang that made it all the way it was. Carroll also points to the idea that "the observable universe is not all there is. It's part of a bigger multiverse. The Big Bang was not the beginning." But even if the Big Bang was the starting point for our universe, research on behalf of electrical engineers at the University of Maryland in early 2011 may have proven without a doubt that time's arrow and entropy do indeed move forward/increase just as predicted, and that time travel as a result would be impossible.

Igor Smolyaninov and Yu-Ju Hung, the two electrical engineers who pioneered the study, reported on by Phys.org in April 2011 ("Modeling of Time With Metamaterials"), used a metamaterial made by patterning plastic strips on a gold substrate, and then illuminating it with a laser. Their idea was to create a Big Bang model and observe the way light moves inside the metamaterial in order to understand how space-time expanded since that rather explosive moment. The experiment's results were consistent with the thermodynamic arrow of time, in which entropy increases within an isolated system, and also examined whether light could move within a metamaterial in a circle identical to particles moving through space-time in Closed Timelike Curves (CTCs), or world lines of particles that make circles and return to their start points. (More on CTCs in Chapter 3.) Unfortunately, within this particular model of the Big Bang, CTCs were proven to be impossible, and that nature seemed resistant to the creation of CTCs, rendering time travel in this mode impossible as well. (But that doesn't mean time travel is impossible in other modes!)

The Entropy Connection

Physicists often ask why the universe also had so much order right afterward, when it could have been a state of utter chaos and disorder. Because of entropy, the forward arrow of time, which increases entropy, the universe instead, was as orderly as it ever got just after the bang. Entropy does not work in reverse, and if it did, it would allow for time to flow in the opposite direction. A hot pot of water eventually gets cold because of entropy, not the other way around. Stephen Hawking, the great physicist, once stated that entropic time is similar to psychological time, or our mental understanding of time, because if the flow of entropy were to reverse, our brain's perception of time would reverse as well. Things around us would become *less* disordered as time passed, instead of what we experience now.

Carroll's questions about entropy, the Multiverse, and the Big Bang also lead us to wonder if the laws of physics might be different in other parts of the universe—or in other universes. Although it still appears that the arrow of time, despite the speed at which it moves, has one direction—forward—and the past is the past and the future is the future, Carroll and others are pushing the envelope in terms of theories and ideas that take our known physical laws and either stretch them or turn them on their heads. One distinction Carroll did make in the article is that even in space where there is nothing to experience time, it still exists.

Time, even if we don't fully understand it, is a fundamental law of nature though it lacks the causality, memory, and progressive forward movement we humans liken to time passing in our own lives. We see time as measurable progress. But that may just be how we see it, not how it really is. Confused now? And you thought time was just a way to gauge when your favorite TV reality show came on the air.

To make matters even *more* confusing, Carlo Rovelli, professor of physics at the Universite de la Mediterranee in Marseille, France, thinks that there is no time! According to an interview for *Forbes* in February 2008, Rovelli suggests that at the most fundamental quantum level there is no distinction between past and future. This distinction disappears in quantum physics. And he believes there is nothing in our known laws of physics that stops this from being the case in the grander, larger sense. "It is only very improbable," he told *Forbes*. "It is only because of our limited view of the world that we reject highly improbable future propositions and turn them into impossibilities." He used the example of a teacup falling to the floor and smashing to pieces. If, as Rovelli believes, we could see past the general picture of the broken teacup and process the detailed location and physical state of each broken fragment, we could predict a future that might seem impossible: the teacup coming back together as a whole and leaping back up on the table.

But this would require a cause, says one of Rovelli's critics, and that causality must be present to make the flow of events move forward from broken teacup to fixed teacup. Rovelli suggests that even causality is a function in the human mind, a notion mirrored in the way philosopher J.M.E. McTaggart described the past as being the future at some point in time, and the future eventually being the past—and that maybe, just maybe, human perception is what drives us to see things the way we do when it comes to the arrow of time.

This Is Your Brain on Time

Without linear time, we would not survive. The ability to perceive, understand, and respond to our surroundings comes from the arrow of time and the lessons we learn in the past that help us to stay alive in the future.

As a child, we may touch a hot stove and get burned. As time goes on, we know that heat and fire can hurt us, and we know not to touch the stove again. Our brains are built to ensure we learn from the past, or, at the very least, remember it.

University of Edinburgh, Scotland, researchers reported in 2008 to *Discover Magazine* that even hummingbirds use linear time-telling to learn just how long it takes for a particular flower to replenish its supply of nectar. Meanwhile, scientists at the University of George learned that rats do the very same thing when it comes to determining exactly when they will be fed.

Neuroscientists at MIT found in a 2009 study led by Professor Ann Graybiel that neurons in the primate brain actually code or stamp time with extreme precision, and that the monkeys can actually then go back and look for the time stamp to recall the event it was associated with when first imprinted. This time stamp and recall are critical for human tasks as well, such as driving a car, playing piano, or keeping track of the events of our lives.

"Soon enough we realized we had cells keeping time, which everyone has wanted to find, but nobody has found them before," Graybiel reported to ScienceDaily.com. These neurons are located in the prefrontal cortex and the striatum, and play critical roles in learning, movement, and thought control.

We have sensory receptors allowing us to touch, see, hear, and smell, but not for time. When it comes to time, our brain must construct them as a type of extra "sense" that is just as important to our functioning and survival.

But when it comes to the perception of time, we can often experience an event, even a past one, as happening more quickly or slowly. All of this is dependent upon what we are experiencing at the time the time stamp is made. For example, a boring afternoon seems to drag on forever, but a fun night on the town goes by in a flash. Time may very well persist as a consequence of the events taking place in it, as philosopher Martin Heidegger once observed. The brain, depending on what we are experiencing, can condense or expand time—not in an objective and real sense, but in a subjective one that the perceiver alone experiences.

Researchers have found in many studies that our judgment of the passing of time may be off, depending on the frequency of stimuli associated with an event, so that relatively infrequent stimuli, such as flashes, pulses, or tones, will increase the speed of the brain's internal "pacemaker." Thus, time that is wasted or frittered away doing "nothing much of importance" often seems to fly by, and we regret how fast we let it do so. In contrast, focusing on goals or challenges that require greater attention will cause the brain to fill the past years with memories, and perceive the passage of time quite differently, as being more richer, longer lasting, and purposeful.

The brain seems to have control over one thing when it comes to time: perception. We may not be able to make a day any longer than 24 hours, but it's up to us how we fill those hours.

In *Time's Arrow and Archimedes' Point*, author Huw Price wrote, "We are creatures in time, and this has a very great effect on how we think about time and the temporal aspects of reality. But here, as elsewhere, it is very difficult to distinguish what is genuinely an aspect of reality from what is a kind of appearance, or artifact, of the particular perspective from which we regard reality." Huw suggested that making this distinction was critical to understanding the asymmetry of time.

So although particular neurons in our brains fire to create time stamps and allow us to recall something we may need later, we still stamp our own perception upon time depending on what we fill it with.

So now the question becomes this: What if our perception of time is wrong?

We see time as motion, as change, as the course or progression of events. But what if time is nothing of the sort?

The truth is, we are still in some kind of infancy stage when it comes to our understanding of time, including how time fits in with space and how our perception of it can possibly shape it.

And yet, we are ever so aware of the clock ticking away the seconds, moments, hours, days, and eventually years of our lives, and we worship time, always wanting more of it or, at the very least, wishing we could go back and re-experience the time we did have.

Maybe we can.

2

TIME IN A BOTTLE

Time is too slow for those who wait, too swift for those who fear, too long for those who grieve, too short for those who rejoice, but for those who love, time is eternity.
—Henry Van Dyke

Imagination is more important than knowledge. For knowledge is limited to all we now know and understand, while imagination embraces the entire world, and all there ever will be to know and understand.
—Albert Einstein

One cannot choose but wonder. Will he ever return? It may be that he swept back into the past, and fell among the blood-drinking, hairy savages of the Age of Unpolished Stone; into the abysses of the Cretaceous Sea; or among the grotesque saurians, the huge reptilian brutes of the Jurassic times. He may even now—if I may use the phrase—be wandering on some plesiosaurus-haunted Oolitic coral reef, or beside the lonely saline lakes of the Triassic Age. Or did he go forward, into one of the nearer ages, in which men are still men, but with the riddles of our own time answered and its wearisome problems solved? Into the manhood of the race: for I, for my own part, cannot think that these latter days of weak experiment, fragmentary theory, and mutual discord are indeed man's culminating time! I say, for my own part. He, I know—for the question had been discussed among us long before the Time Machine was made— thought but cheerlessly of the Advancement of Mankind, and

The original cover of H.G. Wells's The Time Machine.

saw in the growing pile of civilization only a foolish heaping that must inevitably fall back upon and destroy its makers in the end. If that is so, it remains for us to live as though it were not so. But to me the future is still black and blank—is a vast ignorance, lit at a few casual places by the memory of his story. And I have by me, for my comfort, two strange white flowers—shrivelled now, and brown and flat and brittle—to witness that even when mind and strength had gone, gratitude and a mutual tenderness still lived on in the heart of man.

So ends *The Time Machine* by H.G. Wells. Written in 1898, this iconic science fiction story set the stage for the imaginings of many writers and scientists alike, who in their own ways, and for their own purposes, were beginning to open their minds to the idea that we could one day travel back to the past, or forward to the future. In fact, Wells introduced the term *time machine* to readers, and he also was the first to introduce the idea of time as the fourth dimension.

But Wells wasn't the first to touch upon this concept in fictional writings. Long before him, others contemplated what it would be like to move along the landscape of time at will.

Fictional Time Travel

One of the earliest fictional works with time travel as an integral part of the storyline is Samuel Madden's *Memoirs of the Twentieth Century*. Published in 1733, it's the story of a guardian angel from the year 1997 that travels to the year 1728 with letters that describe the future. The full title was originally "Memoirs of the Twentieth Century: Being original letters of state under George the Sixth." British statesman Sir Robert Walpole, a Whig who first served under George I and George II, suppressed the book, now considered incredibly rare.

By the time Wells got around to writing his sci-fi time travel tome about a world filled with peaceful Eloi and malevolent Morlocks, stories of the same ilk had been written by the likes of Washington Irving ("Rip Van Winkle" in 1819), Mark Twain (A *Connecticut Yankee in King Arthur's Court* in 1889, the first time travel fiction in American literature), Charles Dickens (*A Christmas Carol* in 1843), and Edward Page Mitchell ("The Clock That Went Backward" in 1881). Interestingly, it wasn't Wells at all who introduced the concept of an actual time machine. It was Enrique Gaspar y Rimbau who first wrote of an actual machine in the 1887 *El Anacronopete*. The following year, Wells himself would write "The Chronic Argonauts," which featured a time machine made by an inventor.

And after the Wells epic *The Time Machine*, it seemed every writer had to toss his or her hat into the time travel ring. Among the more recognizable names are T.S. Eliot, Robert A. Heinlein, C.S. Lewis, Ray Bradbury, Isaac Asimov, Harlan Ellison, Kurt Vonnegut, Richard Matheson, Dean Koontz, Michael Crichton, J.K. Rowling, and Stephen King! Several authors also wrote sequels and follow-ups to *The Time Machine*, in an attempt to pick up where Wells left off, among them Karl Alexander's *Time After Time*, K.W. Jeter's *Morlock Night*, and David J. Lake's *The Man Who Loved Morlocks*. In 1995, author Stephen Baxter had the privilege of writing the authorized sequel to *The Time Machine*, titled *The Time Ships*.

With the advent of motion pictures and television, the subject of time travel took the imaginings of words and pictures into the world of moving images, resulting in a host of sci fi movies and television shows that dared to wonder about our ability, and inability, to travel the landscape of time. Whether we were going forward into a future we could not yet comprehend but only guess at, or back to a time we knew little about, the fascination has driven an industry that seems to have no saturation point. We want to be entertained, but we also want to *wonder*.

Here's just a sampling of the time travel motion pictures and television shows that have graced our big and small screens through the years:

1960—*The Time Machine*

1964—*The Time Travelers*

1965—*Dr. Who and the Daleks*

1968—*Planet of the Apes*

1972—*Slaughterhouse Five*

1978—*Superman*

1979—*Time After Time*

1980—*The Final Countdown*

1980—*Somewhere in Time*

1981—*Time Bandits*

1982—*Voyagers* TV series

1984—*The Terminator*

1984—*The Philadelphia Experiment*

1985—*Back to the Future*

1985—*Trancers*

1986—*Peggy Sue Got Married*

1986—*Star Trek IV: The Voyage Home*

1986—*Flight of the Navigator*

1989—*Masters of the Universe*

1989—*Quantum Leaps* TV series

1989—*Bill and Ted's Excellent Adventure*

1989—*Millennium*

1992—*Army of Darkness*

1994—*Timecop*

1995—*Twelve Monkeys*

1996—*Dr. Who*

1999—*Austin Powers: The Spy Who Shagged Me*

2001—*Kate and Leopold*

2001—*Time Squad* TV series

2002—*The Time Machine*

2003—*Timeline*

2004—*Harry Potter and the Prisoner of Azkaban*

2004—*The Butterfly Effect*

2004—*Lost* TV series

2006—*The Lake House*

2006—*Click*

2006—*Heroes* TV series

2006—*Torchwood* TV series

2007—*Meet the Robinsons*

2008—*Stargate: Continuum*

2008—*Terminator: The Sarah Connor Chronicles* TV series

2009—*Star Trek*

2009—*The Time Traveler's Wife*

2009—*FlashForward* TV series

2010—*Hot Tub Time Machine*

2011—*Midnight in Paris*

We might ask what came first—the science or the fiction? During the time Wells first wrote of time travel, was the scientific knowledge of that day even thinking of the subject matter? Science fiction has always been forward-thinking, providing readers with a way of dreaming ahead of time, then oftentimes, if one lived long enough, seeing some of those dreams come to fruition in reality. Long before the influences of Tesla, Einstein, and more modern physicists tackling time travel, novelists had to guess at the rate of progress based upon the existing scientific paradigm. The 18th and 19th centuries saw an explosion of scientific advances and discoveries that would give us a better understanding of our world, and inventors were busy creating more and more ways to make life easier. The Industrial Revolution dominated the 18th century, as machines began to replace human and animal "manual" labor. The 19th century ushered in the Age of Enlightenment, when science and rationality began to override the existing focus on religious traditional thought of the times, leading to the American Revolution and the roots of capitalistic thought. Suddenly, we had steel and petroleum products, usable electricity, the advent of steam engines, railways, and steam ships, and the printing press, which allowed the spread of ideas to take hold in a way they never had before.

Inventions That Changed the World

During these two centuries, just a few of the inventions that forever changed the world include:

- 1700s: tuning fork, seed drill, pianos, fire extinguishers, atmospheric steam engine, mercury thermometer, electrical capacitor, lightning rod, sextant, electric telegraph, flush toilet (hurrah!), steamship, submarine, parachute, self-winding clock, steel rollers, threshing devices, safety lock, power loom, torsion balance, guillotine, gas turbine, cotton gin, ball bearings, lithography, batteries.

- 1800s: gas lightning, steam-powered locomotive, first arc lamp, tin can, printing press, first plastic surgery performed, spectroscope, raincoat, soda fountain, stethoscope, matchsticks, Braille print for the blind, typewriter, reaper, electromagnet, stereoscope, ether ice machine, photography, wrench, calculator, propeller, postage stamp, telegraph, revolver, Morse code, hydrogen fuel cell, blueprints, bicycles, stapler, facsimile, sewing machine, rubber tires, antiseptics, safety pin, dishwasher, gyroscope, manned glider, first aircraft engine, fiber optics, internal combustion engine, elevators, pasteurization, plastic, dynamite, torpedoes, windmills, mail order catalogs, air brakes, traffic lights, telephone, moving pictures, practical light bulbs, toilet paper (yay!), seismographs, metal detector, fountain pen, steam turbine, Coca Cola, four-wheeled motor vehicle, radar, contact lenses, gramophone, AC motors and transformers, escalators, zippers, roller coasters, diesel engine, motor driven vacuum cleaners.

These are just a sampling of the products and inventions that came into existence during these two centuries, which culminated in the age of machines, assembly lines, and mass production. But this was also a time when science and scientists gained more respect for their professional endeavors. The word *scientist* was first coined by William Whewell in 1833, who had a way with words, having also coined the words *physicist* and *hypothetico-deductive*. Whewell (1794–1866) was an English scientist, Anglican priest, philosopher, historian, and polymath who took the concept of multiple disciplines of study to a whole new level during a time when most men and women of science focused on a particular discipline. He was the ultimate jack-of-all-trades, but a master of them all as well.

Whewell was deeply involved in the research of everything from physics, mechanics, astronomy, economics, and ocean tides to writing sermons, composing poetry, and eventually writing two profoundly influential volumes that laid out the development of the sciences: *History of the Inductive Sciences* in 1837, and *The Philosophy of the Inductive Sciences, Founded Upon Their History* in 1840. He even had time to come up with his own mathematical equation—the Whewell equation—write extensively about architecture, and expound upon scientific and philosophical values and morals.

Whewell was indicative of the time when science and all it encompassed had begun to move front and center in a world that was previously embroiled in religious thought and debate. During the ages of the industrial and enlightenment "revolutions," science, medicine, and technology were also causing revolutions of their own. Mathematics during the 18th century developed into the "pure mathematics" of algebra, geometry, and calculus and the "mixed mathematics" of optics and mechanics, and in the 19th century ushered in intensified studies in logic, fractal theory, mathematical physics, and theoretical physics, hypercomplex systems, and group theory. At the same time, a second scientific revolution occurred in France with the spread of mathematization, and the Golden Age of physics allowed for the emergence of theoretical physics and the more mechanical paradigm of the coming century. The 1900s would usher in the age of quantum physics, but we will save that for another chapter.

From biology and genetics to applied science, agriculture, and botanical research, Newtonian gravitation, advanced telescopic discoveries (including planets and satellites), radio astronomy, the discovery of radioactivity and the electron, the new science of "industrial chemistry," molecular structure, developmental biology and plant life cycles, geological and geophysical advances in the understanding of strata, the origin of springs, volcanism, tectonic theory, catastrophism, seismography, Ice Age Theory, plate tectonics, oceanography and terrestrial physics, electromagnetism and electricity theories, invention of the battery, natural selection and Mendel's Laws, genetics and molecular biology, thermodynamics and conservation of energy, caloric theory of heat, the Kinetic Theory of Gases, classifications of clouds and precipitation, microbial behavior of germs and viruses, to a variety of stunning advancements in optics, number theory, paleontology, statistics, psychology and medicine, the 18th and 19th centuries offered plenty of brain food for the imaginations of authors and philosophers to feed upon. Even without super-computers and massive telescopes in space and smart phones, the golden ages of advancement opened up a world of possibilities and potentialities that were only dreamed of hundreds of years before.

Now, more than ever, we could not only dream of understanding our world, but even of controlling its mighty forces and natural laws.

Blurring Fact and Fiction

But because science is not the realm of the imaginative or romantic ramblings of science fiction and fantasy writers, the dreams would for awhile stay locked within the realm of "make believe" until enough evidence or data existed to justify taking the extreme ideas of something like time travel into the more academic and professional arenas. Sometimes, though, the lines between the two worlds merged. In fact, H.G. Wells himself was all too aware of what was going on in his world. His fiction was often precognitive of things to come, such as nuclear war, and his understanding of the physics of his day allowed him to imagine the physics of tomorrow, including the concept of time as a fourth dimension. To Wells, much of his fiction involved studying fact and, like any good writer, he did his research and anticipated the progress to come.

In her book *Breaking the Time Barrier: The Race to Build the First Time Machine*, author Jenny Randles refers to Wells's impact on the scientific minds of his time: "The extraordinary possibilities opened up by his novelistic invention of a time machine meant that Wells had created not merely a story, but a new human aspiration that would grow in the minds of everyone who read his novel. That included real inventors and young scientists who would later seek to make his idea a reality."

Randles also documents a real-life attempt by Wells to create a time travel simulator, along with movie producer Robert Paul, in 1895. Their plan was to build a movie theater that would let the audience have the illusion of traveling through time by placing them inside the action via screens projecting the images all around them—sort of like an early IMAX theater. The two couldn't quite make it convincing with the technology of the times, and the plan was abandoned.

The problem, at its root, though, was an indication of a lack of real understanding about time itself. Even as we understood the whole idea of three spatial dimensions of height, width, and depth, because we could see and measure those things for ourselves, time eluded us. Yes, we could measure time, on clocks and calendars, and we could "see" it in a sense that we could see the passing of events, and we held memories of past events as well. But we couldn't find a way to control time the way we could our spatial reality, where we could move about at will and put ourselves in one place or another by choice.

With time, we were stuck in the present, with the past behind us and the future ahead. We were trapped in what appeared to be the linear confines of a dimension that had firm and unmovable boundaries, at least in accordance with the Newtonian physics and laws of nature at the time. In Newton's universe, time traveled as an arrow did, and did not deviate or divert from that straight course. In classic mechanics (also known as Newton's universe), there were physical laws that governed the motion of bodies under actions of forces or systems, and these laws were accurate when applied to large objects. They also applied to anything that could *not* approach or surpass light speed. Isaac Newton's work was based upon earlier works of Galileo, Tycho Brahe, and Johannes Kepler, all of whom created the foundation for the astronomical theories Newton would later built upon.

Laws and Forces

Newton's laws of motion, gravitation, and forces involved everything from velocity, speed, and frames of reference to the forces applied to mass, all the way up to large objects with high velocities that demanded an addendum to the classical mechanics, which came years later in the form of Einstein's theory of special relativity. Einstein's general relativity theory also worked to iron out some of the issues with Newton's law of universal gravitation, bringing physics somewhat closer to that elusive theory of everything (TOE). For years, scientists have searched for one-all encompassing theory that would combine the four fundamental forces of the universe—gravitation, electromagnetism, and the strong and weak nuclear forces—into a one-size-fits-all understanding of reality itself.

These physical laws were backed up by mathematical formulas that did not allow for flexibility in the temporal dimension—not until Einstein came along. As a result, the imaginations of science fiction authors of the 18th and 19th centuries were forced to look far beyond the limitations of Newtonian physics, and many of the machines, devices, and techniques hinted at an understanding that perhaps there existed other laws, other dimensions, and other realities, where time behaved quite differently. Some stories truly stretched the boundaries of the credible with time travel made possible via clocks that can be turned forward or backward, hypnosis, strange portals and doorways, coaches and trains, or, in the case of *A Christmas Carol*, some nosy and intrusive ghosts. In the play *Anno 7603*, written by poet Johan Herman Wessel in 1781, the two main characters are transported through time by a good fairy. In Washington Irving's "Rip Van Winkle," ol' Rip only had to take a nap on a mountain to awaken 20 years in the future. If only it were that easy!

Other modes of time travel in the literature of the times included a very popular method of explaining time travel (keep in mind this was during a time when science itself would not allow for it): having the main character fall asleep, and wake up in the far past or distant future. Wells himself used dreams as a time travel mechanism in *The Dream*, and this served as a popular literary device for those unable to even begin to envision a technological angle. In other stories, the use of special gases, hypnosis, and even being swept overboard into the ocean to awaken a thousand years in the future (the 1824 Russian tale "Plausible Fantasies or a Journey in the 29th Century") allowed for a time travel experience without having to get down and dirty and explain the possible science behind it. If a man fell overboard off a sailboat and woke up in the future, who was to say what might have happened in the water to propel him forward in time? The reader didn't necessarily need to know in order to enjoy the story! Think of the literary devices used today to make a story "work," even as we know in our minds and hearts it is just a bunch of phooey. We buy into it for the time being, because we are being told a story. It was only later that many writers dared to add some solid scientific foundation into their fictional fantasies.

Time Travel Themes and Devices

Whether or not these characters actually did travel, or dreamed it all, was often left up to the readers to figure out, and today this kind of literary device would be considered cheating the reader, kind of like the overuse of the Deus Ex Machina in later fiction and plays. In the 1883 novel *The Diothas*, John Macnie used mesmerism as the means by which his hero was able to visit the 96th century. To get back to his own time, the hero simply fell over a waterfall. How easy is that? But according to Paul J. Nahin, author of *Time Machines: Time Travel in Physics, Metaphysics and Science Fiction*, the sleeping-into-the-future literary device was by far the most popular in the early days of time travel fiction.

Nahin documents many famous and obscure stories that involve sleeping and waking up in the future, citing it as a literary technique that has ancient origins, possibly as far back as AD 600 when Gregory of Tours told a story about the Seven Sleepers of Ephesus, who slept more than 370 years, and the more modern-day Sleeping Beauty, who lay comatose for a hundred years before awakening. The device was most obvious in stories like "Rip Van Winkle," and, as Nahin writes, "This hoary device has been used in modern times, too, to explain how Buck Rogers, born in the last days of the 19th century, could still be around in the year AD 2432." Of course, we cannot forget that Buck was buried alive in a mine cave and kept alive in suspended animation by a strange radioactive gas. When sleeping isn't enough, throw in a cool gas, and you have a recipe for time travel! And if gasses don't strike your fancy, you can try using the freezing method, as utilized in stories like *10,000*

Years in a Block of Ice, or Victor Rousseau's 1917 novel *The Messiah of the Cylinder*, in which the hero goes a hundred years into the future frozen in a vacuum cylinder. Later, in 1956, Robert Heinlein would also revisit the frozen man out of time device in *The Door Into Summer*, which used cold storage to let the hero go into the future and return to the present. As Nahin says, "By 1940, the sleeping-into-the-future technique was so well known that lazy authors hardly bothered to explain it to their readers."

Even horror stories were resorting to the use of drugs as the device that hurtled someone through time, and even H.G. Wells himself fell prey to this intriguing device, in "The New Accelerator," where the professor/hero discovers a drug that causes the person who takes it to experience time running thousands of times faster than normal. The "new accelerator" was the liquid drug itself, and it allowed the world to appear in a "frozen instant," according to Nahin, a theme repeated in many stories to follow by authors who liked the idea of simply ingesting a substance to make time fly. It was just another easy way out in an arsenal of easy ways out that writers often used when they couldn't find the science to back up their stories.

But as the sophistication of science and knowledge grew, so did the expectations of the readers, and the demand for a viable, or at least imaginable, time "machine" put pressure on more and more writers to really do some research and investigate possible ways they might send their protagonists and antagonists into the past or into the future—without resorting to silly things like "mysterious gasses" and falling off a sailboat. Still, they were slaves to the scientific knowledge of their times, and it was only later that authors and filmmakers would up the ante by creating time machines out of DeLoreans and phone booths! But at least they were using material objects!

Anything was up for grabs in terms of a device or trigger or catalyst in the early days of time travel fiction, because it was, after all, fiction. But those authors like Wells who sought to introduce more "realistic" visions of time travel went the extra mile to try to present a model of time machines that at least looked like they could work. In *The Time Machine*, a narrator first introduces a tabletop model of a device that allows a man he refers to as "The Traveler" to travel through the fourth dimension, the dimension of time. That prototype becomes the actual time machine the book's protagonist uses to travel to the year 802,702 AD, where he then encounters the Eloi and Morlocks, and even meets a woman named Weena that he hopes to bring back to the past, or his present, with him. Instead, he travels 30 million years into the future, just in time to see the last gasps of a dying earth.

Luckily our traveler gets back to his lab alive and in one piece, despite having been hurtled time and time again (pun intended) into the vast realms of distant future. What struck readers, and eventually critics, alike was Wells's ability to not just tell a story about future technology, but about the potential for utopian/dystopian societies to arise when technology overtakes humanity. Those who read the book today, even knowing the current cutting-edge advances in science and our understanding of the laws of nature and the universe, can still appreciate the research and attention to detail in Wells's epic.

The time travel fiction of the past was not just about envisioning how far science and technology might one day take us, but also envisioning what life itself might be like. Though this book is mainly concerned with the science behind time travel, we would be remiss to leave out the importance of theme and ideology in the works of these authors, because many of them did touch upon the paradoxes involved in time travel, especially going back to the past and altering it, as well as the possibility of time travel being used for sinister purposes. Some stories even envisioned things like television and how the game of golf might be played, such as J. McCullough's *Golf in the Year 2000*, written in 1892. The author also introduced the concept of equality for women. Other tales, like *Looking Backward*, written by Edward Bellamy in 1888, featured main characters that traveled into the future and woke up in socialist utopias. Others imagined much more bleak, apocalyptic futures, and some even focused on the prevention of historical events by going back to the past to alter the chronology of events themselves, despite the many paradoxes this would create. Dystopic worlds were often the end result of time traveling to the future, thanks to the past and present antics of greedy humans.

Often, characters go back in time to simply observe the past, yet end up changing it, and sometimes they come from the future to help humanity overcome challenges in the present, like Knights and Guardians of Time called upon to change the outcomes of humankind's shared destiny. In his book *Liquid Metal: The Science Fiction Film Reader*, author Sean Redmond writes about the time travel motif as a powerful ideological function that allows the author, and the reader, to

> metaphorically address the most pressing issues and themes that concern people in the present. If the modern world is one where the individuals feel alienated and powerless in the face of bureaucratic structures and corporate monopolies, then time travel suggests that Everyman and Everybody is important to shaping history, to making a real and quantifiable difference in the way the world turns out.

Other time travel stories sought to try to see all the way up to the end of the world, the end of existence, perhaps as a way of finding clues to what leads to the downfall of humanity—and thus avoiding them entirely. Edmond Hamilton wrote in "The Man Who Evolved" of a cosmic ray that allowed the hero to speed up the process of evolution and see how the world ends. In another of Hamilton's stories, *In the World at Dusk*, he has the hero, a scientist, attempt to utilize the past to repopulate a dead future planet, a theme that would become quite common in future science fiction novels and films. The idea that we could either avoid or change a particular fixed outcome is a powerful one. It gives the reader a sense of control over that which is uncontrollable in any other sense: the fate of the world at large.

Later stories would feature lessons about nuclear war and atomic destruction, especially during the 1940s and beyond, and the desire to avoid these calamities or reverse them, should they happen by going into the future to learn a valuable lesson, or going back into the past to stop a horrible choice. In any sense, it was always about control, a theme that resonates today with the popularity of alien invasion films and slasher movies. When things go out of control, what can we do to get control back? Obviously it's a lot easier to do so when you can go back in time, or forward in time, and fix what got broken in the first place.

Maybe it was this desire to be a part of shaping the way the world turned out, and to keep the world from ending, that not only drove writers to imagine time travel were real, but also drove the scientists of the 18th and 19th centuries to push beyond the confines and limitations of Newtonian physics. The desire to understand the universe in which we live in would provide the impetus for scientists, visionaries, and mavericks in their basement labs to push those very boundaries and attempt to conquer time the way we hoped to one day conquer space. Yet we knew that we could travel across the land via foot, horses, carts, and eventually vehicles, and the first hints of air travel and train travel tempted us to look ahead to the day when mankind would break the bounds of terrestrial earth and travel into space. Even if we lacked the technological acumen at the time, the laws of physics allowed for traveling to other planets and our moon.

We *could* do it, though—if we wanted. It was within the realm of scientific possibility. But when it came to time travel, we could only dream of shattering the laws of physics and finding a way to make that dream a reality. Time was different. Time had paradoxes and problems and issues. Time appeared to be linear and fixed; unless we were asleep and dreaming, events happened one after the other, in particular order, and that order could not and would not be deviated from.

Yet we questioned anyway. Would time travel one day be possible? Never did we imagine then that some of the answers to that question might one day come from a world totally invisible to the naked eye. With regard to time travel fiction, author and researcher Jenny Randles discusses the early reactions of scientists to these ideas in her seminal book *Breaking the Time Barrier: The Race to Build the First Time Machine.* Time itself was a concept that had a different meaning to scientists than it did to philosophers. Randles writes that science regarded time as a "fundamental property of the universe (according to Newton) and was controlled by the speed of light at which light conveyed signals...." Philosophers, on the other hand, regarded time as a more subjective experience, as in "time flies when you're having fun." The perception of time, therefore, took center stage and that was something we *could* influence in terms of time travel.

"Time, to scientists, is not open to human intervention in the simplistic manner but is a real property of the cosmos," Randles states. And yet, as she continues, "whether we can or cannot think ourselves into the past, we all do travel into the future with each passing moment of our lives." As Washington Irving wrote about in 1850, we go to sleep and lose all sense of time, and then awaken a day into the future with no recall of the journey itself. Isn't this time travel?

But science demanded physical time travel, involving putting actual material objects, whether a particle or a person, into the future or back to the past. Science, and to some degree the public at large as well, wanted to travel through time physically, not just mentally or in dreams. And but for those pesky laws of classical physics, we could maybe one day do just that, couldn't we?

It would be many years before the universal view of Newtonian physics would be challenged by a whole new realm of scientific understanding that moved attention off the large and cosmic and onto the quantum, where all the laws and boundaries no longer seemed to apply.

3

TIME PASSAGES

Clocks slay time...time is dead as long as it is being clicked off by little wheels; only when the clock stops does time come to life.
—William Faulkner

From the creative ideas and visions of the minds of writers like H.G. Wells, Samuel Madden, and even Mark Twain, who laid down the imaginative framework for the time travel research of the future, came some of the most intriguing and controversial experiments of the 19th and 20th centuries. As we discussed in Chapter 2, many of these imaginers actually understood scientific paradoxes and concepts that would make time travel either impossible, or possible. In 1895, the very same year Wells's legendary novel, *The Time Machine*, was published, Nikola Tesla, considered the greatest unheralded scientific genius ever to walk the earth, had a brush with time travel during experiments with highly charged rotating magnetic fields. Tesla had been researching and experimenting with radio frequencies and how electricity is transmitted through the atmosphere when, according to legend, he accidentally "warped" time and space, and created a doorway through which time travel was possible. Tesla, in fact, has often been referred to as the first time traveler.

Tesla and Time

Tesla was born in Smiljan in war-torn Croatia in 1856. He attended the University of Prague, but later moved to the United States, where he conducted some of the research that put him on par with the likes of Edison. Tesla would invent the AC generator, fluorescent lighting, and other things we now take for granted while under the tutelage of

Edison himself. But Tesla would go on to surpass Edison in achievement, especially when the two became bitter rivals after Tesla proved that AC (alternating current) was far less deadly to work with than DC (direct current), the favored current of Edison. But it was Tesla's more fringe research that put him on the map and his work with EM fields and potential time travel most of all.

In March 1895, after conducting experiments with a huge transformer in upstate New York that produced powerful rotating magnetic fields, which Tesla claimed altered time and space in his immediate environment and made him feel as though he was trapped in a sensation of "timelessness," he began to believe he might actually be able to break down the time barrier. He continued his dangerous experiments, and on one occasion produced a massive ball of 3.5 million volts that detached from the transformer, floated across the room in typical ball lightning fashion, and paralyzed Tesla with a huge charge of electricity that could have killed him. But this contact also pushed him outside of his normal time-space frame of reference. The resonating electromagnetic charge allowed him, even if just for a moment, to glimpse the past, present, and future all at once. Thankfully, his assistant was present at the time and immediately turned off the current, or Tesla would have been fried and charred!

Tesla told this story to a reporter for the *New York Herald* on March 13, 1895, at a café where they met. Tesla supposedly turned up looking quite stunned and out of sorts. He had, after the charge, suffered muscle paralysis and had his nerves jangled, but insisted that he had entered another dimension during that brief incident whereby time behaved much differently than it did here in our three-dimensional existence.

After Tesla's death, physics seemed to slowly come around to actually prove correct some of his findings and validate his research. Many conspiracy-minded researchers also point to his experimentation as the basis for the equally legendary incidences many years later known as the Montauk Project and the Philadelphia Experiment.

Tesla must have known of the dangers involved with messing around with the space-time continuum, and no doubt his direct contact with extreme voltage caused him pause, but it did not stop his research. When he died in 1943, there were many others at the ready to further his ideas and carry on his legacy for groundbreaking research that pushed the edge of the scientific envelope. Tesla's research would be linked to the strange naval military experiment that allegedly occurred in October of 1943 at the Philadelphia Naval Shipyard.

The Philadelphia Conspiracy

Though it has never been proven and remains the fodder of conspiracy theory, the story of the Philadelphia Experiment alleges that a U.S. Navy destroyer escort, USS Eldridge, disappeared—literally. Also known as Project Rainbow, the experiment involved the use of—or perhaps we should say manipulation of—electromagnetism and that, prior to the experiment, the Eldridge had been equipped with devices that would allow for this cloaking to occur. The experiment apparently began in the summer of 1943, with one test in July of that year partially rendering the destroyer invisible and enveloped in a strange greenish mist. Afterward, crew members complained of dizziness and nausea, with all kinds of crazy reports of aftereffects involving sailors having body parts embedded in metal structures and being on decks they were not on when the experiment began.

The general consensus was that perhaps the real motive for the experiment, which the Navy denies ever happened, was to render the destroyer visible on radar, something that made sense in wartime. But according to the legend, in October, the experiment succeeded in doing far more once it was properly adjusted for past mistakes, and this time the Eldridge went AWOL, becoming not only invisible, but, in a flash of blue light, actually transported, or teleported, to a location more than 200 miles away. Supposedly, crew members aboard the SS Andrew Furuseth saw the Eldridge suddenly appear and then vanish again, whereby it reappeared back in the Philadelphia Naval yard. As for time travel, the story is the destroyer went back in time for about 10 seconds.

The origins of this whole story seem to come from a letter sent to Morris K. Jessup, an amateur astronomer and author of *The Case for the UFO*. In 1955, Jessup received a letter from someone by the name of "Carlos Allende," who claimed to be a witness to the antics on the Eldridge as a crew member of the SS Andrew Furuseth. He claimed to have seen the destroyer disappear and reappear, and also claimed to have had knowledge of where some of the crew members ended up. However, Allende had no real backup substantiation for his claims, and, when Jessup pushed for more details, he suspected fraud when new info came under the name of "Carl M. Allen."

Later in 1990 another "former crew member" named Alfred Bielek came forward to support the movie version of the story, *The Philadelphia Experiment*, released in 1984. His version would later be debunked by a team of investigators who concluded that Bielek was nowhere near the ship during the experiment.

Serious researchers contend that this experiment never occurred, and there are indeed a ton of inconsistencies and problems with the various stories presented by those who claimed to be involved. The details can be found in a number of popular books, novels, movies, and even TV shows but, again, nobody really knows what may or may not have happened on that fateful day. From the outcome of the Philadelphia Experiment, many conspiracy researchers believe an even bigger project was undertaken to carry on experiments into everything from psychological warfare to time travel.

The Montauk Project alleges to be a secret undertaking by the United States at the Montauk Air Force Station on Montauk, Long Island. Most accounts claim that the project was a continuation of the Philadelphia Experiment, and the goal was to manipulate the electromagnetic cloak or shielding used on the Eldridge and see what potential military applications it might have. The idea of manipulating the magnetic field, and even gravity, had taken hold of military and scientific researchers alike, and the work being done in secret at Montauk was geared toward developing weapons that could engender psychosis in enemies.

Astrophysicist and UFO researcher Jacques Vallee wrote about the project, claiming that it may have originated from the suppressed memories of a man who was involved in the project, a man named Preston Nichols. Again, the U.S. government and military figures denied involvement, but that didn't stop the conspiracy-minded from labeling the project a Black Ops, supported by the Department of Defense. Some of the wilder claims about the project involved time travel, teleportation, and accessing parallel dimensions, again by manipulation of the magnetic field. Names who have been linked to the Philadelphia Experiment, such as Al Bielek and Duncan Cameron, both who claimed to be aboard the Eldridge when it went AWOL and slipped

Figure 3-1: The Camp Hero radar dish at the notorious Montauk site where alleged time travel experiments once took place.

into a parallel dimension, also discussed their involvement at Montauk, where they claim to have met the famous physicist and mathematician John Von Neumann, even though Von Neumann died in 1957—one of the many inconsistencies both experiments are rife with.

Interestingly, Jacque Vallée once described an experiment that occurred on the USS Engstrom while docked alongside the Eldridge in 1943 involving the generation of powerful electromagnetic field to degauss the ship and hopefully render it invisible to undersea mines and torpedoes. This type of degaussing is actually still in use today, and has no effect on light or radar. Vallée suggests that the real degaussing of the Engstrom may have been at the root of what would later become the legend of the Philadelphia Experiment. Those present at the degaussing, and others who knew of the procedure, might have indeed played a little game of "telephone" and spread rumor, confabulation, and just plain misinformation that resulted in a conspiracy theory that will not go away. In fact, just go on YouTube and search for time machines, and several related to the enigmatic Montauk Project pop up, alleged time machines being created by people obsessed with the idea of utilizing existing technology to achieve such futuristic goals. Of course, none of them work yet, but that doesn't stop these "mavericks" from trying, or from suggesting that their technology has already been stolen by enigmatic government agents and is currently being developed into time travel weaponry!

We explore both the Philadelphia Experiment and Montauk Project, as well as other time travel conspiracies, in Chapter 7 in more detail. But what we find intriguing is the concept, again allegedly tied to Tesla's research years before, of cloaking a three-dimensional object by manipulating the EM field. The idea of cloaking has actually become reality, just as so many of the wild and crazy ideas and theories of science fiction novels and movies of the past often become the facts of tomorrow.

Cloaking Space and Time

In 2006, invisibility cloaking went from being the stuff of conspiracy, conjecture, and science fiction to pure fact when researchers at Duke University's Pratt School of Engineering succeeded in using sophisticated metamaterials to bend electromagnetic waves and partially cloak and render invisible an object. Specifically, the research team, led by Ruopeng Lui, David Smith, and Chunlin Li, created a cloaking device that rerouted certain wavelengths of light, forcing them around objects and making them appear nearly invisible to creatures and machines that see in the microwave spectrum.

The device consisted of a series of concentric circles made of copper rings and wires that were patterned onto sheets of fiberglass. It resembled a loosely coiled reel of film. The device worked only in the microwave range, which meant that the cloaked objects could still be seen by humans, but this was a powerful and profound scientific breakthrough indicating the future possibilities of rendering, with the right metamaterials (artificial materials created to have precisely patterned surfaces that interact with and manipulate light in novel ways), any object visible to the human eye invisible. The ideas behind this breakthrough could be directly traced to the work of Tesla and the possible experimentation aboard the USS Eldridge. Even if that experimentation never occurred the way legend has it, and even if it never occurred at all, the science behind it was, in fact, already being seen as viable.

Sixty-odd years later, it was proved to be just that.

Yet the discoveries continue, and in November 2010, according to CNN's article "Space-Time Cloak Could Conceal Events," new meta-materials promise to control electromagnetic waves in a way that not only cloaks an object in a spatial sense, but in a temporal one as well. Artificial materials designed to manipulate and control EM waves may soon be able to "hide events," according to Professor Martin McCall of Imperial College in London. "In some senses our work is mathematically quite closely related to the idea of invisibility cloaking," he told CNN reporter Simon Hooper. "It's just that we're doing it in space and time instead of just in space."

In a paper published in the *Journal of Optics*, McCall spoke about metamaterials that, upon bending light rays, could also make objects invisible via "blind spots" in time, thus masking a particular event. "If you had someone moving along the corridor, it would appear to a distant observer as if they had relocated simultaneously," he explained, likening it to the famed transporters on *Star Trek*. Another colleague, Alberto Favaro, gave the analogy of someone walking across a freeway full of traffic by speeding up the cars already at or beyond the crossing point, while slowing down the approaching vehicles. Favaro said an observer down the road would see only a steady stream of traffic.

Although the actual metamaterials are decades away from being invented, it is an interesting "theoretical recipe," and one that may certainly work down the road. In the meantime, we have current optical-fibre technology that is already capable of imperfectly hiding events taking place over a few nanoseconds. And lo and behold, in November 2011, scientists at Cornell University announced they had, indeed, created a time cloak that could hide actual events and render them invisible. As reported in the journal *Nature*, researcher Moti Fridman with the School of Applied and Engineering Physics at Cornell and a team of physicists found a way to bend light around an object that made it disappear from view.

This temporal cloaking followed all known laws of physics, yet only lasted 50 trillionths of a second, certainly not enough time to render the discovery usable quite yet, but the implications are staggering. The manipulation of light involved in cloaking is all that is needed, because light carries information. Using man-made metamaterials, the team was able to make light behave in a way it does not naturally behave. The team basically sent a laser beam of green light down a fiber-optic cable and through a lens, splitting the light into two frequencies—one moving faster, the other moving a bit slower. While this was occurring, the team shot a second red laser through the beams.

The beam then entered a part of the cable that allowed the carrying of light of different wavelengths at different speeds, with blue light faster than red. The two colors separated and there was, for a very brief blip of time, no light at all. This blip was referred to by the team as a "time gap" or "time hole," and the beam after the blip was then reassembled by reversing the steps. The time gap was indeed tiny in terms of duration, but it was this time gap that the researchers used to render an event undetected. They did this by pulsing a ray of light through the time gap. Normally this would perturb the first beam in a normal fashion that would be detected when it came out the other end of the cable, but with the time gap in place, when the ray went through and then the beam was reassembled, the detector at the end of the cable showed no change. The gap is actually opened or created by the compression of light, and the experiments indicate that the event occurred, but the detection of it had been manipulated when using detection by ordinary means.

Can the time cloak be adapted to a larger scale? Most researchers involved think that is a long way off, but imagine if it could. The Pentagon's DARPA (Defense Advanced Research Projects Agency) was involved in research suggesting that it may one day have implications on a defense level, perhaps cloaking events on the battlefield during wartime or allowing for covert operations to be carried out in utter secrecy, in fact, invisibility. If we can someday cloak both time and space, we might even be able to render invisible just about anything. The stuff of science fiction novels and movies indeed!

Drouet's Machine

Back to the past we go again, to the year 1949, shortly after the alleged Philadelphia Experiment. The setting is post–World War II Europe, where a French aerospace engineer named Emile Drouet hoped to use the foundation of Einstein's law of relativity for a craft he believed could travel through time. Drouet created a machine, an actual craft, and, with the hopes of finding investors to fund the development of the craft, put it on display in Vigneux-sur-Seine in 1946. This craft, or rocket, was designed based upon Einstein's calculations of the speed of the Earth moving through space. Drouet believed time did not exist except as a result of the motion through space of the Earth. His rocket would travel on a directly opposing path to the tight spiral of the Earth as it moved through the cosmos. Thus, Drouet theorized that his rocket ship could travel back in time one whole year for every spiral of the Earth's path it retraced forward in time.

This rocket was not ever intended to carry a human being, but rather take cameras and measuring equipment with it back in time. Unfortunately, his rocket theory wasn't workable in reality because of the complexity of space-time curves. His spiral theory was too simple and too basic—but held within it the seed of an idea that might actually work one day, if the spirals were not actual physical movements, but mathematical patterns. With further research, Drouet's theory could be attemptable, but with a device that would have to travel faster than 150,000 mph and be much more complex than what the French engineer had ever imagined. Drouet found no further financial interest or backing, but he did become a media darling for a while, even if the scientific community at the time chose to give him no further attention. He was, perhaps, the first real scientist to succeed in, at the very least, trying to come up with a device or design for a workable time machine—even if that machine's workability still needed, well, a lot of work.

And in Drouet's favor, the very same year he sought money for a full-scale version of his time machine, German mathematician Kurt Gödel (1906–1978) actually came up with a mathematically viable and more sophisticated rendering of Drouet's idea. This time, the scientific community took notice. Gödel, who had worked with Einstein while at Princeton at the Institute of Advanced Study, worked with the math to formulate the theory that time travel into the past was possible, if the universe itself was rotating. Like Drouet's spirals, one could hitch a ride on a time machine craft that rode the rotational paths of the universe, which Gödel called "closed time-like curves." One would still be traveling below the speed of light, but the circular or spiral-like path would allow one to go back to the past and then eventually come back to the present.

Gödel's Time Machine

Figure 3-2: Gödel almost got it right.

In 1951, Gödel demonstrated the existence of paradoxical solutions to Albert Einstein's field equations in general relativity, offering his discoveries to Einstein as a 70th birthday gift! Gödel's "rotating universes" would allow time travel and indeed caused Einstein to have doubts about his theory. Gödel's "Metric," as it became known, involved closed timelike curves, the subject of much controversy in physics, and was also known as the Gödel's Solution because of the strange properties it embraced that did allow for a form of time travel. Unfortunately, this solution required properties, such as a rotating universe, that just weren't a reality at the time.

Depressing as it was to realize that our own universe did not rotate in such a way that Gödel's theory would work, it drove home the point that time travel was absolutely possible once we figured out the math and how it applied to our particular universe! Critics of Gödel moved in on the paradoxes presented by the time travel aspect of the Gödel Solution, which theoretically would allow for travel into the past. The bases of their arguments were usually associated with the resulting paradoxes of a time traveler that, in Gödel's universe would end up meeting himself in the past—his own past. The rotating universe model, his critics suggested, was flawed right out of the box simply because it included time travel as a possibility. Physicists would later argue that Gödel's universe, if finite, would remove the time travel aspect completely, making everyone more comfortable with the theory, which remained just that: theory.

So, big deal. The closed timelike curves didn't work here (they could very well work in other universes, though!). All it would take was some tweaking of the theory to apply to the known laws of physics in our universe and, with a little more research, ingenuity, and brainpower, we would have a time travel theory that would become a time travel fact. Perhaps Gödel's epic fail wasn't epic after all, inspiring a whole new breed of scientists researching the possibility of time travel, and how to finally get it right and overcome those nasty paradoxes. (More on those in Chapter 5.)

Though today scientists at CERN's Large Hadron Collider, in the foothills of the Jura Mountains outside of Geneva, Switzerland, collide particles via a massive underground tunnel, their hopes are that we will soon find proof of parallel universes, which would allow for potentially new and even unimaginable laws of physics to be present that might make it possible for an object to travel through time, even if the laws in our own universe do not yet allow for it. In Chapter 6 we will look at the amazing discoveries at CERN so far, and what dreams may come in the next year. There are even whispers of finding evidence of extra dimensions of both space and time.

That's the way it works with ideas being born in the minds of great thinkers and dreamers. One day, those ideas find an outlet and become physical, tangible—real. In Chapter 4, we'll look at some of those fictional ideas—tunnels through space, surpassing the speed of light, bending space-time, alternate dimensions, additional temporal dimensions, wormholes and shortcuts through time, time dilation, and more—and see how many of them have come true. Surprisingly, it's more than you might have ever imagined. Travel forward into the next chapter with us and find out.

4

TIME AFTER TIME

Our heirs, whatever or whoever they may be, will explore space and time to degrees that we cannot currently fathom. They will create new melodies in the music of time. There are infinite harmonies to be explored.
—Clifford Pickover, *Time: A Traveler's Guide*

I would not want to bet against the possibility of time travel. My opponent might have seen the future and know the answer.
—Stephen Hawking

In 1971, a physicist and an astronomer teamed up to test Albert Einstein's theory of relativity in a more practical manner. Their experiment involved using four cesium-beam atomic clocks. The physicist, Joseph C. Hafele, and the astronomer, Richard E. Keating, had these clocks mounted on board commercial airplanes that flew twice around the world, one time eastward, the next westward. The clocks were then compared with the United States Naval Observatory's famed atomic clocks, which are said to be the most accurate on earth.

The result was published in Volume 177 of *Science* in 1972 in an article titled "Around the World Atomic Clocks: Predictive Relativistic Time Gains." The experiment was not done to test atomic clock accuracy so much as to prove that time is indeed different depending on where the clock was and where it was headed. The Hafele-Keating Experiment showed that, consistent with the special and general laws of relativity, there were differences in observable time losses and gains relative to the frame of reference. Here's an excerpt from the article:

During October, 1971, four cesium atomic beam clocks were flown on regularly scheduled commercial jet flights around the world twice, once eastward and once westward, to test Einstein's theory of relativity with macroscopic clocks. From the actual flight paths of each trip, the theory predicted that the flying clocks, compared with reference clocks at the U.S. Naval Observatory, should have lost 40+/-23 nanoseconds during the eastward trip and should have gained 275+/-21 nanoseconds during the westward trip.... Relative to the atomic time scale of the U.S. Naval Observatory, the flying clocks lost 59+/-10 nanoseconds during the eastward trip and gained 273+/-7 nanosecond during the westward trip, where the errors are the corresponding standard deviations. These results provide an unambiguous empirical resolution of the famous clock "paradox" with macroscopic clocks.

Interestingly, today's GPS systems had to take these slight time adjustments into consideration when satellite mapping locations on Earth. Modern technology, it seems, has to constantly keep up with the times!

Relatively Speaking...

Albert Einstein's theories of general and special relativity and time dilation set the foundation for the potential of time travel now being heavily researched today. Those concepts we all learned about in grade-school science still present a template by which we judge our ability to one day take a jaunt into the past, or hurtle forward into the future. Einstein, more than any other scientists of his time or before him, has influenced time travel and even the study of time itself. And in his own words, well, it's all relative, and it has a lot to do with light and curves. Curved space, that is.

Relativity theory basically informs us that motion is not an absolute and that objects move in relation to one another. What an observer at rest sees differs from what an observer in space—moving—sees. In fact, the observer at rest would seem to be the one in motion to the one who actually *is* in motion out in space. Bring the speed of light into the mix, and it gets even more confusing.

Say you are in a spaceship that is at rest, sitting in space going nowhere. A moving spaceship whips past you near the speed of light, which is 186,000 miles a second or 700 million miles an hour. The guy in the moving spaceship will think *you* are the one moving, even though you are stationary. And to the guy in the moving ship, time will slow down and, if he doesn't have anything at all to reference, he won't even know *he* is actually moving.

It's all relative. Actually, to give props where props are due, this is often referred to as Galileo's Dictum, because even before Einstein perfected the laws of relativity, Galileo had been doing some relative thinking of his own. Galileo had his own relativity principle way back in the 17th century, suggesting that people on Earth could not distinguish between being at rest or moving with the rotation of the Earth each day. His thought experiment of the time involved a cannonball falling from the top of a ship's mast. Whether or not the ship is moving at sea or docked at rest, the cannonball will land upon the base of the mast—*and* the people on board the ship observing this cannonball trick would not be able to tell if they themselves are at sea or stuck at dock without an exterior reference. If they are just looking at the cannonball land on the mast, they won't be able to tell what the hell the ship is doing! In other words, the only way for a person to determine if the ship is in motion or at rest would be in relation to the exterior environment. The ship would have to be moving relative to something else!

Special Theory of Relativity

In 1905, Einstein published his special theory of relativity, which gave time itself a bit of flexibility. Clocks that travel along different routes will measure different elapsed intervals of time. Einstein also incorporated the speed of light as a constant for all inert observers, despite the state of motion of the object they were observing. And, nothing could ever go faster than light—period. In fact, his Principle of Invariant Light Speed stated that "light is always propagated in empty space with a definite velocity, or speed, independent of the state of motion of the emitting body."

Special relativity, which incorporated the works of Hendrik Lorentz and Henri Poincare, as well as the foundations set by Galileo, only dealt with relativity in terms of frames or objects of reference that were in relative motion in respect to one another. But this theory left out the effects of gravity, which prompted Einstein to develop the general theory of relativity. This took into account the curvature of space-time and Newton's law of universal gravitation, eventually identifying gravity and its effects as a geometric property called space-time, or the curvature of space-time.

In Einstein's 1905 paper, "On the Electrodynamics of Moving Bodies," he summed it up for us rather simply by stating the two postulates of special relativity that we are most concerned with when it comes to time:

1. The laws of physics will always be the same for all observers that are in uniform motion relative to one another.

2. The speed of light in a vacuum will always be the same for all observers, regardless of their relative motion and regardless of the motion of the source of light itself.

As the fourth dimension, time behaves like the three spatial dimensions in special relativity. Time expands as speed increases. Length will also contract as speed increases. This only applies when moving forward, not back. In spatial dimensions, we can move backward. In the temporal dimension, it seems, we cannot.

Space Wars!

When it comes to theoretical physics and mathematics, there are two kinds of space. ***Euclidian space,*** named for Greek mathematician Euclid of Alexandria, has only three spatial dimensions, and none of time. Three coordinates can determine each point in the 3D Euclidean space, one for each dimension of height, width, and length.

Add in the fourth dimension of time and you get ***Minkowski space,*** named after mathematician Hermann Minkowski, who suggested that space and time should be treated equally, with events taking place in a unified 4D space-time continuum. In 1908, Minkowski wrote in *Space and Time* that "space by itself, and time by itself, are doomed to face away into mere shadows, and only a kind of union of the two will preserve an independent reality." To this day, we continue to think of space-time as unified. However, there are theoretical physicists such as Itzhak Bars of USC College, whose research was the focus of a May 2007 article in PhysOrg.com, who suggest there might be a second temporal dimension that could unify Einstein's theory of gravity and quantum theory. It would also require an extra dimension of space, though. Bars posits that our 4D world might be the shadow of a richer six-dimensional reality, with the reality we see as the world around us like a two-dimensional wall displaying shadows of objects in a three-dimensional room. This mirrors the Holographic Universe theory that posits that our reality might be a three-dimensional image projected onto a two-dimensional surface from possibly a higher dimensional origin point.

Time Dilation

These theories further led to the concept of time dilation, which posits that a moving clock will tick more slowly than the stationary clock of an inert observer. General relativity posits that clocks run slower in deeper gravitational wells, called gravitational time dilation. Interestingly, general relativity also posits that the universe itself is moving away from us, the inert observers on Earth, faster than the speed of light, which is contradictory to the finite nature of light speed postulated in the special theory of relativity. Special relativity, after all, states that any particle that has rest mass, the minimum value allowable of the relativistic mass of a particle at rest, can never be accelerated fast enough to equal or surpass the speed of light. To put it simply, time dilation may seem surreal, but it isn't. It is simply describing the relativistic relationship between space and time. Time to the resting observer will appear as constant and absolute. It's when movement comes into play that we experience the effects of relativity.

With time dilation, the flow of time will also slow near a gravitating body. And although time will expand from the perspective of the observer at rest, space contracts from the perspective of an observer in motion. The closer you get to the speed of light, the stronger the time dilation effect will be, and it is negligible at the normal speeds of, say, a car or plane or train. Time, then, seems to be directly attached to the perception of the observer, dependent upon his or her frame of reference, which led German philosopher Immanual Kant to suggest that time and space are both properties of the mind and perception, and that we project these perceptions onto the natural world.

The Michelson-Morley Connection

In 1887, two scientists teamed up to perform what would become one of the most famous scientific experiments: the Michelson-Morley Experiment. Albert Michelson, born in 1852, was an American physicist who spent years perfecting the measurement of light speed (which would help him win the Nobel Prize for Physics in 1907). Edward Morley, born in 1838, was an American scientist noted for his work in optics and physics. The two men joined forces and worked for several years on their now-famous experiment to prove the existence of "luminiferous aether" in which electromagnetic waves propagated. The idea was that light waves would require a medium to move through, just as sound waves used the mediums of air and water. This would apply even to a vacuum, because light could indeed travel through a vacuum. Thus, proving the existence of the medium of light was a main goal of their experiment. *That* part of the experiment was a huge failure, but the two men did succeed in solidifying that the speed of light does not vary depending on direction of measurement or the position of the earth's orbit. The speed of light as a constant for all inertial frames of reference (in the absence of gravity) became the foundation for Einstein's special relativity theory. Recall that two lights flashing simultaneously will appear to flash at the same exact time to observers at rest, yet will appear to flash at intervals to an observer or observers in motion. Objects at rest see the lights flash at the same time even if they are hundreds of miles away from each other.

In September 2009, a team of physicists from Germany did their own updated version of the famous Michelson-Morley Experiment and again confirmed that the speed of light is the same in all directions. The team, led by Stephan Schiller, performed their measurements at the Heinrich-Heine University in Dusseldorf, and it has been called the most precise experiment to date. The team hopes to increase the precision enough to possibly detect dark energy one day as a violation of the Lorentz Symmetry, which many physicists hint may not continue to stand the test of time as even more highly precise measurements are undertaken and we learn more about how gravity may be compatible with quantum physics.

For now, the idea that there is an exact symmetry of nature, and that light speed is the same in all directions, remains the law of the land—or, well, the law of the light.

Time dilation, as permitted in the special relativity theory, does, in fact, suggest that forward time travel is doable. Time appears to be going by more slowly for faster-moving bodies, relative to an observer that is stationary, and the closer that body gets to the speed of light, the more time slows. A clock would appear to literally stop working, with its hands frozen in time, so to speak. This is only a type of time travel into the future, and not into the past, that for many paradoxical reasons we will explore in Chapter 5.

Mass and Speed

The problem arises when we consider a crazy little thing called mass. Einstein's relativity theory tells us that objects gain mass as they accelerate to greater speeds. In order to get something moving at all, though, you need a force or push to move it. The more mass an object has, the bigger force or push you will need, and as that object approaches light speed, it will also approach infinite mass, meaning—you guessed it—it will need infinite push, or acceleration, or it stops moving. As of the writing of this book, we do not have any kind of time machine or starship that is able to provide infinite push!

Time dilation isn't just something that holds for everyday speeds, though, as a recent experiment by James Chin-When Chou and colleagues at the National Institute of Standards and Technology in Boulder, Colorado, discovered. In October 2011, Chou's findings were reported on PhysicsToday.org showing that time dilation was present at a rate of 10 m/s or 22 miles/hr using two of the world's most accurate clocks. The experiment wasn't front-page news, but it does support special relativity's assumptions. However, there is a call for the same experimentation using non-atomic clocks to possibly disprove special relativity, something suggested by Einstein himself before his death.

Despite all of these theories, we are seeing potential breaches of what has long stood as fact in the cutting-edge research of today. Right now, relative to Einstein's theories (yes, pun intended), would time travel be possible? Without going into the deep physics involved, which would take up two books itself (see the Bibliography for suggested reading), we can take the theories of relativity, time dilation, and light speed limitations, and approach the concept of time travel by first asking: If light speed is unattainable, and any traveler in space-time will always be moving slower than light, how can we ever stop the arrow of time from moving forward into the future?

In July 2011, physicist Joao Magueijo attempted to find proof that the speed of light hasn't always been constant and that, in the earliest stages of the existence of our universe, light traveled faster. The 40-year-old native of Portugal, former faculty member at Princeton and Cambridge, and author of *Faster Than the Speed of Light*, posits that the speed of light is a "dynamic variable" that may prove to be the bust for Einstein's constancy of light speed theory. Magueijo suggests that light speed may congeal to zero near a black hole and be much higher in the presence of cosmic strings, which would permit high-speed travel without time dilation effects. His continued research may end up proving that in the past, at least, light wasn't confined to the speed we apply to it today.

Did We Mention the Twins?

One of the most interesting aspects of time dilation involves a pair of twins. Let's call them Joe and Jay. Joe decides to take a trip into space to check out the big Wal-Mart built on a nearby planet 25 light years away. Jay isn't feeling well, so he stays home and watches the San Diego Chargers beat the crap out of the Denver Broncos. Jay, by the way, is a Broncos fan. Joe travels in his little rocket ship at very close to the speed of light—let's say 99.99 percent close. The trip to Wal-Mart and back takes 50 years, give or take a day or so just to find parking. But when Joe gets back to Earth, he pops in on Jay and is shocked to find that Jay has aged 50 years and is now an ornery old guy. Joe, on the other hand, is only six months older than he was when he left, and still fairly attractive and kind of temperament. That's because Joe's trip only lasted *six months* for him out in space, but on Earth, it came out to *50 years* of passing time.

But worry not, for Joe didn't get an extra 50-year pass on aging. He literally, in space, only *lived* six months, while his twin brother Jay lived the full 50 years on Earth—because it's all relative!

For now, let's assume light speed cannot be breached so readily. Is time travel possible in Einstein's world? In Chapter 3, we looked at one possibility, of closed timelike curves, and the research and work of Kurt Gödel, but for now we want to look at a theory Einstein himself proposed.

It's a bridge of sorts—through time and space.

Bridges and Wormholes

You would have to be living under a rock not to know what a wormhole is. Television shows and movies galore promote the idea of a shortcut through space and time, such as the traversable wormhole in *Star Trek: Deep Space Nine*, which allows the characters to travel over vast spatial distances. Sometimes wormholes can even get you from one time in space to another. Remember the rabbit hole in Lewis Carroll's *Alice in Wonderland*? The author based it upon the wormhole research of German mathematician George Friedrich Bernhard Riemann, who is considered the first scientist to really tackle the concept.

Born in 1826, Riemann exhibited amazing mathematical skills from an early age and spent his school years immersed in a love for math that would lead him to found an entire field of geometry, known as Riemannian Geometry or the Riemannian Metric. His work would become a foundational element of Einstein's later general theory of relativity, and would also lead him to develop a wormhole theory of connections between spaces with zero length, called Riemann Cuts. At the time, Riemann did not assume his theory was applicable to traversing between universes via a wormhole, but his ideas of connected spaces would also imply higher dimensions (he is credited with formulating higher-dimensional geometry) and, as always in science, prompted further research by the likes of German mathematicians Felix Klein and Adolf Hurwitz, and even Einstein.

Riemann Cuts were shortcuts. In the cosmic sense, they were shortcuts that connected two unrelated regions in the space-time continuum, and presented a faster way to get from one point to another, known as "multiplied connected space." The term *wormhole* wouldn't be formally introduced until 1957, courtesy of American theoretical physicist John Archibald Wheeler, who referred in his *Annals of Physics* to the topology of multiply connected space that Riemann Cuts supported, but a solid theory itself took shape around 1921 when another German mathematician, Hermann Klaus Hugo Weyl, proposed his theory of the connection with mass analysis of electromagnetic field energy. Weyl heavily studied Einstein's general relativity theory, and he eventually published a series of papers that focused on the theory's physical applications. He wrote a book titled *Raum-Zeit Materie (Space-Time Matter)* in 1918, detailing his comprehensive analytical work of the geometric aspects of relativity theory and how they relate to space-time physics, and is even credited as being the driving force behind today's modern gauge theories in physics. Even the potential discovery of the God Particle, or Higgs Boson, which may be the missing link in what gives mass to matter (more on this later), is dependent upon Weyl's prior work into the mathematical symmetry of gauge invariance. Weyl, in fact, represents a number of lesser-known brilliant minds who helped develop and tweak the theories of those who came before them and laid the groundwork for scientists yet to come.

Traversable Wormholes

Einstein would further the wormhole dialogue with his collaborative work with Nathan Rosen, a physicist who also took part in the famed EPR Paradox with Einstein and Boris Podolsky, questioning the entangled wave function of quantum physics and the whole "spooky action at a distance" argument of particles communicating with one another over time and space. Rosen and Einstein teamed in 1935 when he became Einstein's assistant at the Institute for Advanced Studies in Princeton, New Jersey. Together, they discovered the mathematics behind one type of wormhole that could connect different points in space, involving a black hole at one end and a white hole (a black hole that moves backward in time) at the other. This wormhole became known as both the Einstein-Rosen Bridge and a Lorentzian Schwarzschild Wormhole, named for German physicist Karl Schwarzschild, who

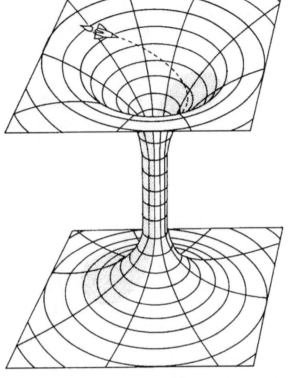

*Figure 4–1: A Lorenztian (Schwarschild)
wormhole. A black hole serves as the entry
point, with a white hole the exit point.*

provided the first exact solution to Einstein's field equations of general relativity, just weeks before Einstein himself released his theory to the public. Schwarzschild's solution eventually led to the Schwarzschild Radius, which defines the size of an event horizon of a non-rotating black hole. The geometry of the Schwarzschild solution involves again a black hole, a white hole, and two universes that are connected to one another via their horizons by a wormhole.

The black hole would be the entry point, with the white hole as the exit point, and with the good ole' Einstein-Rosen Bridge as the wormhole that bridges one to the other. But it isn't that easy when it comes to actually *traversing* a traversable wormhole. First of all, white holes appear to violate the second law of thermodynamics (entropy).

Also, there's this thing called singularity, which is the point at which anything, including light, entering a black hole will meet with infinite gravity and be crushed or stretched into complete utter nothingness. The center of each black hole has this singularity, this point of infinite space-time curvature and incomprehensible gravitational effect, and because even light cannot survive, the only possibility of surviving entry into a black hole would be to travel beyond the speed of light.

Then there's the issue of holding the "throat" of the black hole entrance long enough to get anything through it, although many physicists point to exotic matter as a possibility (see later in the chapter). You also have to make damn sure the white hole exit point is wide open when you get there, if you get there in one piece.

When developing a theory, there are always challenges, obstacles, and problems left for the next generation of brilliant minds. The same is said for wormholes. If the Einstein-Rosen Bridge were nothing but a theoretical oddity at the time, and an example ultimately of a non-traversable wormhole, what would make them traversable?

First of all, if traversable, a wormhole could then be categorized into two different types: *intra-universal* (the wormhole would connect two points in space or time in the *same* universe, and *inter-universal* (the wormhole would connect two points in space or time in *different* universes).

The Stability Problem

Before we go into parallel universes and the Multiverse, we need to first be sure we can travel through a wormhole at all. In 1962, physicists John Wheeler and Robert W. Fuller presented a research paper that proved the Einstein-Rosen Bridge unstable, even for light, to get from one exterior point to the other exterior point. Then, in 1963, New Zealand mathematician Roy Kerr took the debate to the next level by providing a newer more detailed description of black holes that might actually overcome some of those obstacles. Kerr made the assumption that a star that would collapse into a black hole would be rotating and, instead of eventually collapsing to a point or singularity, would instead collapse into a ring. His concept came from the study of neutron stars, which are massive collapsed stars with the mass of the Earth's sun. Thus, as Kerr posited, a dying star would collapse into a rotating ring of neutron stars. The centrifugal force would then keep that nasty singularity from occurring. Though the amount of gravity would be huge, an object could make it through the wormhole and come out the other side. This new type of black hole was known as a Kerr Ring, or Kerr Black Hole. It exists in theory, until we find the technology to traverse such a ring.

■■■■■

Time Tunnels

Fiction and film have long been a stomping ground for combining potential scientific theory with just plain fun story-telling. When it comes to black holes and wormholes as potential time travel portals in film and television, we've been blessed with the good, the bad, and the downright ridiculous. We all remember the sci fi Disney epic *The Black Hole*, which broke new special effects ground when released in 1979 and introduced audiences to the wonders of these cosmic gravity suckers. Some of the more memorable recent uses of tunnels through time since then include:

- *Terra Nova* (2011)—The mega-budget sci fi series, executive produced by Steven Spielberg, sent families back in time 85 million years to the dinosaur days. Why? Because the Earth of 2149 could no longer sustain the environmental damage we humans unleashed. The mode of transport: a supercharged star gate that you simply walked through, that attached the gray, gloomy present to the wild, wild past.

- *Fringe* (2010)—This hugely successful sci fi television series features the character of Dr. Walter Bishop opening a wormhole into an alternate universe that is prominent to the ongoing plotline.

- *Star Trek* (2009)—Although many episodes of *Star Trek*— the television series as well as the movies—featured shortcuts through space and time, this film was memorable for Spock's use of a material called "red matter" to create an artificial black hole. The characters use the black hole to go back in time and change the past.

- *Donnie Darko* (2001)—This cult movie fave offered a portal through time when the doomsday-vision-riddled lead character, played by Jake Gyllenhaal, used a wormhole to transport a jet on the day of impending apocalypse 28 days back into the past.

- *Star Gate SG-1* and spin-offs (1997–2007)—The hugely popular sci fi series, based upon the 1994 movie *Stargate*, featured wormholes and black holes in many episodes as portals through space and time.

- *Andromeda* (2000)—This television sci fi series featured a black hole that allowed the hero to survive by slowing time.

- *Farscape* (1999)—This Australian-produced sci fi series involved a character that comes through a wormhole and ends up in a distant universe, and is then involved with the race to build wormhole technology to get him back home, although others hope to use the technology for more sinister purposes.

- *Contact* (1997)—Based on Carl Sagan's novel of the same name, this film used a series of wormholes utilized by a crew of five humans, Jodie Foster included, to make a trip to the center of the Milky Way. There, Jodie's character met with her long-dead father, then came back to meet a resistant scientific inquiry and a pastor who looked an awful lot like Matthew McConaughey!

- *Event Horizon* (1997)—This feature film was able to travel faster than the speed of light and passes through an artificial black hole into another dimension.

- *Sliders* (1995)—A group of time travelers uses a wormhole to "slide" between parallel universes in this popular television sci fi series.

- *Babylon 5/Crusade* (1993)—This American sci fi "space opera" television series featured "jump points, " which were artificially created wormholes that allowed faster than light travel and hyperspace access.

This is just a tiny sampling of the many films and television shows that use black holes and wormholes as portals through time. We were going to mention one of the most memorable at all, *Bill and Ted's Excellent Adventure*, a 1989 classic about two slackers who travel through time using a wormhole, but thought you might lose all respect for us as authors!

Kip Thorne's Wormhole Theory

Another traversable wormhole theory involved keeping the throat open with some type of exotic matter or matter with negative mass/ energy. This became the theory du jour around 1985, when cosmologist Kip Thorne set out to come up with a hypothetical traversable wormhole. It was the popular astronomer and astrophysicist Carl Sagan who asked Thorne to do so, by the way. Sagan was working on his novel *Contact* at the time and needed a way to get his heroine, Eleanor Arroway (played by Jodie Foster in the movie version) to Vega, a star located 26 light years from Earth. Thorne came up with the idea of sending "Ellie" through a wormhole to cut the otherwise approximately 490,000-year trip (were Eleanor's craft to travel at the speed of the fastest spacecraft at the time, Voyager) down to a much shorter time period. The wormhole idea worked smashingly for the film, and helped propel Thorne into the media limelight as the "man who invented time travel," when he published his wormhole theory in the year 1988 in a physics journal.

Though his idea was not yet technologically feasible, Thorne did also propose a wormhole that would not create any time paradoxes, or kill anyone inside of it, using the Casimir effect, in which an electrical field is created between two parallel metal plates. The Casimir effect proved that negative energy densities do occur in nature, and perhaps this negative energy density could be the magic, "exotic" potion needed to stabilize the throat of a wormhole long enough to travel through.

Thorne, also a maverick and groundbreaker in the world of black hole research, worked with many fellow physicists to develop various wormhole theories, including Mike Morris and Ulvi Yurtsever. The three published their research in the journal *Physical Review Letters* and described how an advanced civilization might create a large wormhole, stabilize it to prevent it from collapsing, and convert it into a time machine that could transport an object or information/communication back and forth in time. The paper presented a challenge for all wormhole research by presenting a very science fiction–sounding idea with good science behind it, yet opened the discussion and debate as to how their wormhole might actually one day be more than theory.

Faster Than Light!

Traveling through a wormhole could also avoid the whole light speed barrier thing. Wormholes such as the ones Thorne proposed would allow for superluminal travel because FTL, faster than light relative speed, would apply only locally. By ensuring that the speed of light is not breached locally at any given time, you could theoretically send something through a wormhole faster than it would take a light beam to cross the same path *outside* of the wormhole. For the person inside, he or she would be traveling at subluminal speed, but would still beat the light beam on the outside. Not so, though, if the light beam also decided to travel through the wormhole. It would then be faster!

Wormholes had been theorized to exist on a quantum level, in such tiny scales as part of the quantum foam, blipping in and out of existence, never existing long enough to possibly inflate to a large enough scale to allow for a human to fit through. However, some physicists have suggested these microscopic wormholes as a possible way to transmit information back in time, if not an actual object. Most physicists agree that any speculation about ways of sending even information back in time must first clear the hurdle of coming up with a single theory that combines relativity and quantum mechanics. Until then, it's all just fun and guessing games.

Hawking's Response

Stephen Hawking, the brilliant theoretical physicist, scientist celebrity, and author of the popular *A Brief History of Time*, *The Universe in a Nutshell*, and *The Grand Design* books that popularized physics for the lay audience, was at odds with Thorne's wormhole theory. Hawking declared that the entrance to any wormhole would emit so much radiation that it would be unstable, and even closed off permanently. (He also shot down physicist Frank Tippler's 1974 concept of a Tipler Cylinder of infinite length, rotating fast enough to produce closed timelike curves, and thus, time travel. Hawking's 1992 paper, "Chronology Protection Conjecture," in the *Physical Review* stated that closed timelike curves could not be created. Thus, no time travel. (Oh well!) One of Hawking's own theories involved parallel universes connected by wormholes, like bubble universes connected to each other, but tiny and allowable of only quantum leaps between worlds, rather than on a grander scale.

Stephen Hawking has been asked about time travel innumerable times, and in May 2010 he wrote a piece for the *Daily Mail* called "How to Build a Time Machine." It was subtitled, amusingly enough, "All you need is a wormhole, the Large Hadron Collider, or a rocket that goes really, really fast." In the essay, Hawking discussed the dreamer in him that drives his scientific research, despite his suffering from ALS, also known as Lou Gehrig's disease, a disease of the nerve cells in the brain and spinal cord that control voluntary muscle movement. The disease took his body (he has been wheelchair-bound for years), but not his amazingly sharp and inquisitive brain.

Hawking pondered the possibility of time travel within the known laws of nature, and he looked at wormholes as that possibility in the *Daily Mail* piece: "The truth is, wormholes are all around us, only they're too small to see. Wormholes are very tiny. They occur in nooks and crannies in space and time." He went on to explain that it is a basic physical principle that nothing is solid or flat, and that there are holes and wrinkles everywhere—even in time. Crinkles, voids, and crevices in time, he called them, referring to the smallest scale of the quantum foam. Tiny tunnels that provided shortcuts through space and time, although they didn't stick around for long, popping in and out of the foam.

Hawking's wormhole universe allows for the theoretical possibility of a giant wormhole constructed in space, but he doesn't find it feasible, mainly because of the paradoxes and the problem of causality, sort of temporally putting the cart before the horse. "That kind of time machine would violate a fundamental rule that governs the entire universe—that causes happen before effects, and never the other way around. I believe things can't make themselves impossible," he wrote.

The problem goes back to wormholes, and the fact that as soon as one expands, natural radiation enters and it all ends up in a loop. The wormhole is then destroyed. Even the idea of inflating a tiny wormhole would meet with this same end. But, Hawking only pointed to these issues as obstacles for time travel into the past. He does believe time travel into the future is possible, pointing to supermassive black holes and time dilation. The supermassive black hole would act as a time machine itself. This is not practical, and Hawking set his sights on another way to travel through time.

It would require traveling incredibly fast, and with the current limitation of light speed, that speed would be 186,000 miles per second. No faster. Traveling near this speed will transport you into the future, relative to anyone who is *note* doing so. The closer you get to light speed, the more time slows, in much the way that short-lived particles called pi-mesons have done in the Large Hadron Collider at CERN, in Switzerland, where tiny particles are sent on a journey around a 16-mile-long circular tunnel until they smash into each other at incredibly high speeds.

Hawking's well-known argument against time travel, though, asks why, if there is such a possibility, we are not surrounded by time travelers from the future. Carl Sagan, in an interview for PBS.org, stated that he finds this argument dubious, just as he finds the argument that aliens don't exist just because they aren't everywhere on Earth: "I can think half a dozen ways in which we could not be awash in time travelers, and still time travel is possible.... First of all, it might be that you can build a time machine to go into the future, but not into the past, and we don't know about it because we haven't yet invented that time machine." Sagan also suggested that time travel into the past *is* possible, but that the time travelers from the future haven't gotten to our time frame yet. And, perhaps time travel back in time is possible, but only up to the moment time travel itself is invented!

Sagan also suggested that time travelers may already be here, but we can't see them—maybe because of superior technology, or invisibility cloaks, perhaps. He even referred to ghosts and aliens and fairies as potential time travelers. His final argument against Hawking's argument *against* time travel was that perhaps we just weren't advanced technologically enough yet to master it: "I'm sure there are other possibilities as well but if you just think of that range of possibilities, I don't think the fact that we're not obviously being visited by time travelers shows that time travel is impossible." Sagan also pointed to wormholes as a possible mechanism for bridging two points in space and time.

Our understanding of the universe we live in changes daily, with amazing new research and cutting-edge discoveries challenging what we thought was fact and law. In a two-month period in 2010 alone, PhysOrg.com and Space.com reported some truly stunning research being done by those in search of answers. In March, a theoretical physicist from Indiana University, Nikodem Poplawski, published in the journal *Physics Letters* his theory that our universe exists within a larger universe, based upon his wormhole research involving the radial motion through the event horizon of two different types of black holes, the Schwarzschild and the Einstein-Rosen. His paper posited that all astrophysical black holes might have Einstein-Rosen Bridges, or wormholes, each with a new universe inside that formed simultaneously with the black hole. "From that it follows that our universe could have itself formed from inside a black hole existing inside another universe," he stated in the journal. Both Phys.Org and Space.com excitedly presented the findings of Poplawski's research, which suggests again that our universe was born in a black hole. He felt that his theory might also explain what happened before the Big Bang and whether or not our universe was closed.

His theory also presented an unusual angle to the entire wormhole/exotic matter debate. All wormholes are unstable, in principle, and close off the instant they open, and the suggestion that exotic matter could keep them open continues on until we actually prove or disprove the existence of this exotic matter. Poplawski suggested though that perhaps when black holes form, matter undergoes a phase transition to become exotic matter, thus allowing the initial expansion of the universe inside the black hole.

Out of the Mouthes of Babes?

In January 2010, CNN and other news outlets reported an amazing story about a 13-year-old boy named Gentill Abdulla, who proposed his own theory of time travel involving a time machine, based on his research into the "areas of black holes, time travel, wormholes, magnetism, light, and most important, gravity," according to CNN.com Abdulla had devised an experiment that theoretically, he claimed, could allow time travel. This experiment involved using magnets in front of each other and a beam of blue light that had been traveling for hundreds of thousands of years. You put the beam of light between the magnets, then turn on lasers on each magnet, and all kinds of stuff would happen with the magnets that would eventually become smaller than its Schwarzschild radius and become black holes. Okay, so it was a really complicated experiment that, we suspect, has not been proven to result in a real time machine yet, because we could find no follow-up articles or reports, the sheer ingenuity and intellect of this child makes us wonder if time travel of the brain is possible and if these theories will indeed one day become fact. All from the mind of a 13-year-old who could have been wasting his time playing Halo: Reach or shooting Nerf guns....

By the way, this boy is not to be confused with 12-year-old genius Jacob Barnett of Indiana who was reported in *Time*'s March 2011 issue to be close to disproving Einstein's theory of relativity with a series of mathematical equations he has come up with. Professors at the Institute for Advanced Study in Princeton, New Jersey, have confirmed he's on the right track to coming up with something completely new. For now, they're encouraging Barnett to continue doing what he likes to do, which is explaining calculus using a whiteboard marker and his living room windows, some of which is posted on YouTube. Just Google it!

With an IQ of 170, we are hoping Barnett will also help Abdulla come up with a workable time machine before either of them turns 16. Perhaps the technology we need already exists, in the minds of the future? Stay tuned!

In Chapter 6, we'll further delve into the most current research involving time travel science and the men and women at the forefront, or behind it, whichever time frame you prefer. But first we ask: What truly stands in the way of time travel being achieved with the level of knowledge and technology we have today—as of this moment? Speed of light limitations, laws of physics as barriers to time travel, and, of course, those lovely paradoxes.

What are these paradoxes all about, and can they ever be overcome and eliminated from the scene all together? Let's go ask our grandfather.

5

TIME WON'T LET ME

People assume that time is a strict progression of cause to effect, but actually from a non-linear, non-subjective viewpoint—it's more like a big ball of wibbly wobbly...time-y wimey...stuff.
—Steven Moffat

The past is obdurate.
—Stephen King, *11/22/63*

Time travel...will never be impossible forever.
—Toba Beta, *Betelgeuse Incident*

If you could choose one seminal moment in history to return to and relive what would it be? Would it be the birth and death of Jesus? The building of the pyramids in Egypt? The signing of the Declaration of Independence? Perhaps you would want to stop the assassination of President Abraham Lincoln or John F. Kennedy? Maybe you simply want to relive the birth or death of a close family member?

It is more than likely that each of us has a different and unique answer, but it is a universal certainty that we have a particular moment in time that we are drawn to, perhaps a formative event in our own lives that we continually return to in our mind's eye?

Now that we have identified the time and place, the next most important question is probably the most difficult: *How?* How can we actually traverse the grid and move at will throughout the time-space continuum? Is it a simple act of getting in our customized DeLorean, engaging the flux capacitor, and putting the pedal to the medal until we reach 88 miles per hour? Because Doc Brown and Marty McFly did exactly that in *Back to the Future*, it must be true, right? Unfortunately not. However, as we shall see in later chapters, scientists, researchers, authors, and hobbyists have proposed a variety of possible mechanisms and theories regarding legitimate time travel. Physical time travel? Mental? Astral?

If time travel is possible, we must also seek out barriers or restrictions that would thwart evildoers from utilizing it for illicit, illegal, or immoral purposes even as we address the currently known laws of nature or physics that would prevent one from traveling forward or backward in time. Special relativity theory tells us that nothing can go faster than the speed of light, and that, if it did go faster, it would move backward in time. This suggests, then, that in the cause and effect arrow of time reality we live in, the effect would come before the cause. Imagine: a door opening before you push it open. Your waiter bringing you a rare steak exactly the way you like it before he took your actual order. A sentence appearing on your computer screen before you type the keystrokes. It turns everything we know to be a part of the order of our world on its head. But is it possible?

Time Travel? No Way!

In July 2011, a research team led by Professor Du Shengwang from the Hong Kong University of Science and Technology claimed that they had empirically proven that a single photon obeys Einstein's theory that nothing can travel faster than the speed of light. If peer review ends up proving the veracity of their claim, then this may throw a fairly significant kink in the plans of aspiring time travelers, as it demonstrates that, outside of science fiction, physical time travel is impossible—at least within the constraints of our physical laws of nature.

On its Website, the university stated, "Einstein claimed that the speed of light was the traffic law of the universe or in simple language, nothing can travel faster than light. Professor Shengwang's study demonstrates that a single photon, the fundamental quanta of light, also obeys the traffic law of the universe just like classical EM (electromagnetic) waves."

In January 2001, physicists discovered that superluminal (faster than light) propagation of optical pulses in certain specific medium might be possible. This highly controversial finding reopened the possibility of time travel. However, it was later determined that the phenomenon could be explained as simply a visual effect.

Throughout the debate, Professor Shengwang maintained that Einstein was correct and theorized that by measuring the ultimate speed of a single photon (which had never before been done) scientists would know with certainty whether or not superluminal propagation was a viable possibility. "The study, which showed that single photons also obey the speed limit c, confirms Einstein's causality; that is, an effect cannot occur before its cause," the university said. Shengwang added, "By showing that single photons cannot travel faster than the speed of light, our results bring a closure to the debate on the true speed of information carried by a single photon."

Though fascinating, this experiment has been met with some healthy skepticism. Some physicists believe that their research seems to only demonstrate the actual speed of photons. Others believe that this only proves that the speed of light cannot be broken by known particles. Perhaps there are particles (as yet undiscovered) that can travel faster than light. Because most scientists unequivocally state that they don't know everything about the universe, could we also assume there are laws of physics yet to be discovered? We will get to that in Chapter 6, but for now let's assume this limitation still stands.

Paradoxes

Beyond the limitations imposed by our current understanding of physics with regard to physical time travel, are there any other theories that might forever relegate time travel to the purview of science fiction? Yes, several possible paradoxes have been theorized. Among the more popular ideas are the following:

- Grandfather paradox.
- Chronology protection conjecture.
- Bootstrap paradox (Ontological paradox).
- Predestination paradox.
- Restricted action paradox.
- "You" overpopulation paradox.

Grandfather Paradox

The most widely known and quoted paradox of time travel is the grandfather paradox. The paradox, mentioned in the 1943 book *Le Voyeageur Imprudent* by Rene Barjaval (although he was not referring to a grandfather in the story!), posits first of all that you are able to build a physical, functioning time machine and go back in time. Then you decide to travel back in time to meet your grandfather before he meets your grandmother and produces any children (your father or mother). Now, suppose you go back, meet him, and decide that he's not the cool grandpa that you thought. In fact, he is evil and vicious. You return back to your time line and, after imbibing in a few too many adult beverages, hatch an irrational plan to once more go back in time—but this time to kill him. According to the grandfather paradox, if you did so, you would never have been born, and the time machine would not have been built. The grandfather paradox is a protection mechanism that would kick in and prevent you from doing so, thereby averting a "loop" in time. In short, the grandfather paradox provides protection from changing the past. A time traveler would be able to view the past; however, he would be unable to interact with it.

Interestingly, in the movie *Back to the Future*, the lead character, Marty McFly, comes face to face with an opposite sort of grandfather paradox: In the movie, he must get his father to meet, fall in love with, and marry his mother in order for him to be born! This paradox, although the most prevalently discussed, is also one that theoretically might be the easiest to overcome, based up on new ideas about the Multiverse and alternate time lines (which we will get into more in Chapter 6). Hypothetically, if you go back and change an element of the past that would extinguish your own birth, you would, theoretically, not exist in this present—but there are ways around this if we accept the possibility that just because *this* past time line cannot be altered, another time line can't be.

An offshoot of this theory is the Non-Existence Theory, which states that if you went back to the past and changed anything that would cause you to not exist (such as stopped your parents from meeting), such as "auto infanticide," which is going back and killing yourself as an infant, you would end up with a sort of *It's a Wonderful Life* or *Scrooge* world where you, and all of the causes and effects that would have otherwise been associated with you, never happened. In fact, *It's a Wonderful Life* is a great example of a variation of this paradox that manages to demonstrate in the most entertaining of manners that, if you do something in the past that would bring about your own non-existence, once you went back to the future, you would be alive, but living in a world void of any effects of your past actions. So think about George Bailey seeing his beloved Mary as a poor and bitter spinster librarian because she and he never hooked up! She didn't even recognize him, and was in fact afraid of him! Poor George Bailey. If we took the time to think hard about all that we have caused and effected over our lifetime, we can see how our non-existence not only changes *our* past but the pasts of others—countless others.

In early 2011, an experiment was conducted and reported in Physical Review Letters ("Closed Timelike Curves via Postselection: Theory and Experimental Test of Consistency") that would avoid the grandfather paradox. This theory involved CTCs, *closed timelike curves*, which are paths in space-time that return to their starting points. Led by MIT researcher Seth Lloyd, a team that included scientists from the Scuola Normale Superiore in Pisa, Italy; the University of Pavia in Pavia, Italy; the University of Toronto; and the Tokyo Institute of Technology conducted an experiment involving CTCs that behave like ideal quantum channels similar to those involving teleportation. These CTCs, Lloyd and his team pointed out, are self-consistent and post-selected, and work by projecting out part of the quantum state. "In normal physics (without closed timelike curves) one specifies the state of a system in the past, and the laws of physics then tell how that state evolves in the future. In the presence of CTCs, this prescription breaks down: the state in the past plus the laws of physics no longer suffice to specify the state in the future," Lloyd stated. The experiment, using photons, also involved setting up an initial condition, then simulating how the system would evolve in the future as final conditions the scientists hoped to impose as their post-selected outcome. "Whenever that outcome occurs, then everything that happened in the experiment up to that point is exactly the same as if the photon had gone backward in time and tried to kill its former self. So when we post-select that outcome, the experiment is equivalent to a real CTC." The scientists used two qubits in a single photon, one of which represented forward-traveling and the other backward-traveling to demonstrate their theory, and recognize that even the existence of CTCs are theoretical, resulting in the need for more research.

Though the grandfather paradox remains the most prominently known obstacle to traveling into the past, physicist and author Thomas Roman suggests in his book with retired physics professor Allen Everett, *Time Travel and Warp Drives*, that there are other ways around this paradox. When interviewed for the December 2011 Slate.com blog, "Time Travel: Beyond the Science Fiction," Roman suggested two ways to avoid this paradox. One would be if you went back to the past and attempted to kill your grandfather, but something prevented you from doing so and, thus, you didn't do the one act that would then prevent your own existence. "The other possibility is you go back in time, kill your grandfather, and at that point the universe splits. The murder of your grandfather occurs in a different universe." The universe you are in, in which you did kill your grandfather, is the universe in which you were never born, so you have no past, only a future history. "You live out your life in that universe for the rest of your life, but you can't get back to the universe that you started from." Your past would be wiped out of existence along with your grandfather.

Carl Sagan put it this way, in his PBS.org Nova interview about time travel, "Sagan on Time Travel." If you travel into the past and murder your grandfather before he sires your father or your mother:

> Do you instantly pop out of existence because you were never made? Or are you in a new causality scheme in which, since you are there you are there, and the events in the future leading to your adult life are now very different? The heart of the paradox is the apparent existence of you, the murderer of your own grandfather, when the act of murdering your own grandfather eliminates the possibility of you ever coming into existence.

Sagan posits that perhaps you just plain cannot murder your own grandfather. You might shoot him, but the gun jams or he dodges the bullet. Nature itself may somehow contrive to prevent the very act that would interrupt the causality scheme leading to your existence.

Chronology Protection Conjecture

The chronological protection conjecture is a theory proposed in 1992 by physicist Stephen Hawking that states that the laws of the universe are constructed in such as way that time travel is prohibited. In Hawking's conjecture, the hidden laws of the universe will protect it from a time paradox by not allowing a time machine to be successfully constructed and utilized. As an example, if one were successful in creating a time travel device that utilized say a wormhole traversal, these universal laws would kick in to stop its use by destroying the wormhole with electromagnetic vacuum fluctuations that feed back on themselves through the wormhole.

Remember Kip Thorne's wormhole at the request of Carl Sagan? Thorne's theory was a real "thorn" in Stephen Hawking's side because of this chronology protection conjecture, which Hawking came up with as a way to discount the possibility of time travel simply because, well, it wasn't possible. The laws of physics wouldn't allow for it (although recall from the last chapter that even Hawking admitted this wasn't as set in stone as he had first intended, at least for time travel into the future). But Hawking's conjecture does allow for minor time travel, at the sub-microscopic level, where it cannot substantially affect anything on a grander scale.

The absence of time travel again may be linked to the law of causality, and our universe is set up to not allow for breaches of that law. In fact, this may be linked to the Anthropic Principle, in that our universe is especially designed with the evolution of intelligence in mind and that time travel could never itself evolve because it would make impossible the ability of a living thing to experience events in a future-driven fashion, make predictions and assumptions that lead to survival and adaptation, and even basically understand the world by being able to place and sort events into a chronological order that describes the experience of life itself. The fundamental principle of causality may remain the biggest time travel thorn to contend with, because the violation of said law seems impossible as of yet even at the quantum level, and the universe may follow this unbreakable rule to the letter, both on a cosmic scale and on a quantum scale.

Bootstrap Paradox (Ontological Paradox)

The bootstrap paradox takes on the question of information and time, stating that an object or information can exist without first having been created! According to scientists, this concept is opposite of the grandfather paradox in that the information or object sent back in time becomes or creates its past self. The term refers to "pulling yourself up by your bootstraps," an expression made popular by Robert Heinlein in his story "By His Bootstraps." The bootstrap paradox is a favorite of science fiction writers, and several fictional stories are based on this paradox. One example is in the American drama TV series *Lost*. On *Lost*, Richard Alpert gives the character named John Locke a compass in the year 2007. Locke is later sent back approximately 60 years into the past, where he gives the compass back to Alpert and tells him to bring it back to him in 2007. In this particular example, the compass itself is the paradox.

Many people live by the "whatever happens was meant to happen" mentality, which is the bootstrap paradox in action. No matter how many times you go back in the past and change something, nothing changes, because that is how it was meant to be. It's like a loop that cannot be escaped, because any change will still result in the same outcome. In Heinlein's story, the bootstrap paradox comes into play when the protagonist goes through a time portal after a stranger requests it, and then the protagonist is met by a second stranger who tries to stop him from going through the portal. The three men get into a fight, and the protagonist is pushed through the portal—only all three end up being the protagonist himself. The first stranger is his future self, and the second an even farther future self who wants to prevent the loop from occurring by keeping the present self from entering the portal.

Predestination Paradox

The predestination paradox (also called a closed time loop) is commonly referred to in science fiction movies and books and is very similar to the bootstrap paradox (begging the question, why do we need so many paradoxes?). Basically, the predestination paradox involves the creation of a causal loop when the time traveler gets caught in a loop of events that predestines his or her travel back in. A time traveler who went back to the past to change or alter history would then find himself fulfilling his role in the creation of "normal" history, as we know it, and not actually changing it. It also implies that if a time traveler exists in the past, she was in the past before just by her presence in her *now*.

In the same sense that you might have a destiny to fulfill in life, going back to the past would mean you don't change your destiny; you just fulfill it as it was always planned that you would! Here's a great example: A young woman goes back in time to find out why her horse escaped from the barn. While in the past, she falls asleep while scrubbing out the barn, leaving the stable door open, causing her horse to make for the hills, and creating her desire in the future to want to go back to the past and lock the damn lock. Got it? She fulfilled her own destiny all over again!

This paradox can also involve information and sending information back in time that might affect someone in the future. An example of this would be going to a psychic, who tells you that you are going to get hit by a bus on Tuesday. So on Tuesday, you stay at home to avoid traffic, only to have an errant bus driver crash through your living room wall with his Greyhound and break every major bone in your body. Again, the main crux of this theory is the sense of a self-fulfilling prophecy, where even going back in time to possibly change something results in the very thing happening just the way it played out in the present.

On a deeper level, what if the universe itself has a destiny? It's a valid question, and there are physicists at work trying to determine if it indeed does, such as Paul Davies of Arizona State University in Tempe and physicist Yakir Aharonov, whose research into the paradoxes of quantum mechanics inspired him. Davies is working to try to understand the very possibility that there is a set destiny to our universe and that the future may somehow (perhaps on a quantum level) be reaching back into the past to affect the present. This would completely unseat our views about causality and time itself, positing that the fate of the universe is like a hand stretching far back into the distant past and influencing it. If so, what influence would we, the time travelers, have on our own influence upon the past, or the future? If it's all set in cosmic stone, then our role may be as travelers in the present moment only.

Restricted Action Paradox

In the restricted action paradox, some physicists believe that the laws of nature (or perhaps some other, unknown intervening cause) would activate and somehow forbid the time traveler from taking any action that could later result in his or her own time travel from occurring. In this paradox, the time traveler simply cannot change history. The restricted action paradox and the grandfather paradox share many similarities—the most obvious being the inability to manipulate the past time line. This paradox is also associated with the Novikov Self-Consistency Principle, developed by Russian physicist Igor Dmitriyevich Novikov in the mid-1980s, which asserts that if any event exists that would give rise to a paradox, or to any "change" to the past whatsoever, then the probability of that event happening is zero. Time travel is therefore impossible. Enough said.

Novikov posited that nature will only allow for behaviors and actions that are self-consistent, and therefore nature would not allow for a paradox to arise. A self-consistent solution was always present and always made the most sense. What has happened in the past cannot be "unhappened" (just like that old saying "A done bun can't be undone") and is considered to be a done deal. It can't be changed or altered, and it cannot be repeated twice in two different ways. Just as the laws of physics our universe operates by restrict certain actions, these same laws would restrict a time traveler who wants to go back and change the past. The laws are immutable.

This puts forth the impossibility of a time traveler doing something in the past that would then prevent the time traveler from later traveling back in time! So say you go back and try to shoot your grandfather, but the gun jams and your grandfather escapes, thus protecting the future from any change to the past. Time travelers to the past, therefore, cannot change history, and, as the predestination paradox states, often a time traveler in a science fiction scenario will intend to change the past to prevent a particular action, only to find that his or her attempt actually precipitated the action!

So perhaps you go back in time to shoot Hitler before he can commit the atrocities of the Holocaust, only to fail to stop Hitler and, in some way, actually allow him to carry out said atrocities. Your intentions, honorable as they were, would simply hold no water because nothing you did in the past could change the past.

The "You" Overpopulation Paradox

This paradox is a bit hard to follow, however, but fun to think about, so please try to stay with the rest of the class! Imagine jumping into a time machine and traveling back to yesterday. In yesterday, there are now effectively two of you (the you from the future, and the you from now.) At the same time, there is also a you approaching the time at the present when you first got into the time machine—about to get into the time machine yourself. That you then enters the time machine and travels back to a world where there is no longer one you but, now, two yous. Actually now there are three—and soon a whole endless string of you's approaching the time machine at the present about to travel back to yesterday. Make sense? Basically, this paradox is theorized to provide protection against multiple you's from existing in parallel worlds (even though each is from a different timeline). Doppelgangers, anyone?

These paradoxes all seem way more like philosophical impossibilities than truly scientific ones, but the limitations they impose are mind-twisting. Could you change the past? An ad for a Stephen King novel states: If you go back and fix the past, make sure you don't break the future. That novel, *11/22/63*, released in November 2011, is a perfect modern take on the question of messing around with the past, and is highly recommended for anyone who likes to see how fiction explores time travel and the associated paradoxes. In the novel, set in the present day, the lead character, Jake Epping, finds a local diner that connects him with the year 1958 and is able to go back and forth in time. We won't give any spoilers, but his main goal is to go back and figure out a way to stop Lee Harvey Oswald from assassinating John F. Kennedy on November 22, 1963. Simple enough, and Epping even decides to right a few other wrongs when he goes back—only, from the start, disturbing paradoxes arise that make for some mind-blowing reading. His desire is to save certain individuals from unspeakable crimes, but in doing so he realizes that he may have overstepped his temporal boundaries and caused even more suffering, and that undoing the damage becomes more and more difficult with each trip back to the past. Apparently, if you *do* try to fix the past after breaking the future, your window of opportunity gets smaller and smaller.

Ultimately—and, trust us, the book is worth digging into—Epping learns that by going back to fix the past, you might actually be making the future worse, no matter how good your intentions might be. Even when he attempts to *not* make any changes on his trips back, his slightest decisions are almost impossible to duplicate exactly as they were in the previous trip, resulting in unwanted alterations to the time line. The moral of the story is that your actions in the past, even the smallest and most inconsequential of them, have untold effects on an untold number of people in the future, bringing the Butterfly Effect into play.

The Butterfly Effect

The Butterfly Effect is a part of chaos theory that states that even the smallest change at one location or state in a nonlinear system can result in a larger change to a later location or state. Edward Norton Lorenz, an American meteorologist and mathematician, considered a pioneer of chaos theory, first presented his ideas of "sensitive dependence on initial conditions" in a 1963 paper for the New York Academy of Sciences in the context of meteorological research into weather prediction. In the paper, he wrote that a fellow meteorologist stated that if the theory were correct, it would imply that the flap of a seagull's wings would change the course of weather forever. His colleagues encouraged him to change the seagull to a butterfly, and the rest, as they say, is history.

The idea is that the simple flapping of a butterfly's wings, which seems totally inconsequential, actually affects the atmosphere (albeit in the most minute ways at first) and could theoretically then cause a hurricane thousands of miles away because of this sensitive dependence upon the initial conditions—the initial condition here being a pretty butterfly doing its thing and flying from flower to flower. In terms of time travel paradoxes, and as King's novel so succinctly describes, the effects of even the smallest action can be massive, and were we to go back in time not to make a huge change, such as killing Hitler before he could kill millions, but something simple as kissing the girl we weren't brave enough to kiss at the prom, could still result in untold and completely unwanted changes to a number of people's lives. Do we have that right? Does anyone have that right?

To alter the time lines of oneself alone is hard enough to do without screwing up the time lines of everyone we come in contact with, everyone we know, and everyone who knows someone who knows us.

It's that chain of cause and effect that our mere existence sets off. Just stop reading right now and think about all the people you know, and have come in any kind of contact with over the course of your lifetime, and try to imagine how even the simplest of greetings or comments, as well as your most intense involvements, might have affected others. It's almost impossible to imagine how far our reach goes, like a pebble tossed into a pond. Our ripples actually never stop, because each person we touch and effect is then a carrier of our ripple, added onto their own.

The flap of your symbolic wings could cause a figurative hurricane in the lives of people thousands of miles away, and you might not even be aware of it. So going back in time and making a tiny tweak might not be the non-event you think it is.

Another, more humorous example of this concept can be found in the notorious *Family Guy* cartoon series, specifically in an episode titled "Back to the Pilot," which was the fifth episode of the 10th season— and aired, quite coincidentally, the same month Stephen King's time travel novel was released! In this episode, baby Stewie and Brian the talking dog use a time machine to go back in time to the very first episode of the show to find a lost tennis ball. Once back to the date the ball was believed lost, January 31, 1999, the two come upon the Griffin family and begin noticing all kinds of changes to the past that were not the way they recalled in the future. Brian also tells his past self about the September 11th terrorist attacks, and this sets off a Butterfly Effect of changes that end up resulting in the present day being altered completely and quite negatively.

When Stewie and Brian return to the present time line, they realize their actions of changing the past and preventing the September 11th attacks made a completely apocalyptic mess of the future. Of course, this calls for a return to the past to undo the changes they made on the first trip back and return the future to its original course, but, like the King novel, it drives home the disturbing point that though we may long to go back and fix things in the past we think would have prevented suffering, either to ourselves or to billions of people, doing so might in fact make things worse. Much worse.

One of the questions that arises from this seeming impossibility to change the past is the idea of free will. Do we, if everything is on a fixed time line that we cannot alter, have any choice? Do we truly possess any free will? If the outcomes are all predetermined, and we are living in a universe of unchangeable constants, then where does the idea of free will enter the picture? Perhaps we are only able to exercise free will in the present, and because each moment of our lives in really only experienced in the present, we do have all the free will we could ever want. But once it is done, it is done.

So, even with the odds seemingly against physical time travel (especially with all of the possible paradoxes), let's suppose that physical time travel is, in fact, achievable (and many theoretical physicists still do acknowledge the possibility!). Why, then, do most scientists agree that *if* we can time travel, it can only be forward—and not backward in time? To answer this, we need to go back to the speed of light itself.

Most scientists would agree that light travels at a speed of 186,282 miles per second. As we detailed in Chapter 4, according to most conventional interpretations of Einstein's theory of special relativity, due to the time-slowing effect that physical matter seems to experience as the speed of light is approached, movement through time is believed to stop at the very moment that something attains "c" (the symbolic designation for the speed of light). Based on this, many have come to speculate that moving faster than light might possibly equate to traveling backward through time. Of course, there is no evidence of this; however, it certainly would seem to make sense.

Recall Professor Shengwang's research at the Hong Kong University of Science and Technology conducted in July 2011 showing that light speed does appear to be a barrier. This seminal research not only seems to confirm causality, but also seemingly rules out the possibility of backward time travel.

What about time itself? In physics, time is described as a dimension similar to length, width, or height. Every day when you travel to work or home, not only are you traveling through a physical direction in space, but you are also moving forward in time—the fourth dimension. Charles Liu, an astrophysicist with the City University of New York, College of Staten Island, and co-author of the book *One Universe: At Home in the Cosmos*, says, "Space and time are tangled together in a sort of a four-dimension fabric called space-time." Liu explains that space-time can be thought of as a form of spandex encompassing four dimensions: "When something that has mass—you or I, an object, a planet, or any star—sits in that piece of four-dimensional spandex, it causes it to create a dimple. That dimple is a manifestation of space-time bending to accommodate this mass."

Does the concept of space-time itself imply that space and time are somehow conclusively linked? According to the known physical laws of nature (and again, it is simply what we know at this point in time), the bending of space-time causes objects to move on a curved path. This curvature is referred to as gravity. Mathematically, we can go backward or forward in the three physical spatial dimensions, but time does not share the same multi-directional freedom. According to Liu, "In this four-dimensional space-time, you're only able to move forward in time."

Perhaps the question we need to be asking is not regarding whether faster than light speed (FTL) is necessary for time travel, but "what is the speed of time?"

Throughout the history of mankind, the application of science has helped to expand the knowledge of our physical world. Recall some of the more well-known scientific "facts" that were later disproven, such as the Earth being flat, cold fusion, Einstein's static universe, the Phlogiston Theory, the Expanding Earth—and countless others. Give it some time and perhaps science will once again correct itself.

Born This Way?

We would like to add just one other time travel paradox. We call it "the Lady GaGa paradox," and it simply states that maybe, just maybe, we can't travel in time because we weren't "Born This Way." The human body is built for living within the confines and limitations of the three spatial dimensions and the forward-moving arrow of time. It serves us. It's the way it is. Until we develop brains and bodies that allow us to travel in time, it won't matter what goes on in the cosmic world, or the quantum world, and even the coolest of time machine devices will fail to overcome this Occam's Razor of paradoxes. It just ain't meant to be for us right now. It's the simplest of all explanations and the easiest paradox to swallow.

As kids, we all wanted to be birds and fly. Many of us even attempted flight by jumping off of coffee tables and beds, flapping our arms, and breaking a wrist or ankle when we clumsily hit the ground. Yes, we can fly in planes and jets, but *we* cannot fly and we cannot turn ourselves into birds, or horses, or butterflies. We just weren't "born this way." We like to think of ourselves as supreme, and able to overcome any challenge or obstacle, but we keep forgetting we are human and relegated to live as humans, with all the rules and limits that apply.

So love or hate Ms. GaGa, you have to admit she was onto something....

And yet we still look to the future when we can overcome those rules and limits, do we not? All of these paradoxes involve the known laws, and even those that are only theoretical, of this universe that we call home. Perhaps, then, the way around them is to think outside of the box—or at the very least, outside of this universe, where maybe it doesn't matter if we were "born this way" at all!

Time circuits on...Flux Capacitor...fluxing...Engine running...All right!
—Marty McFly, *Back to the Future* (1985)

6

TIME HAS COME TODAY

There comes a time when the mind takes a higher plane of knowledge but can never prove how it got there.
—Albert Einstein

In 2013, millions of loyal fans and devotees will celebrate the 50th anniversary of what has to be the definitive time travel television series: *Dr. Who*. Originating in 1963, this Guinness Book record-breaking series is the longest-running science fiction television series of all time, and, for those addicted to it, the most successful as well.

Dr. Who focuses on the adventures of an intelligent and eccentric humanoid alien on the run from the Time Lords of the planet Gallifrey, his origin planet, and his trusty companions who travel about space and time in a device called the TARDIS (Time and Relative Dimension in Space), which takes on the outer appearance of a blue phone booth or police box, thanks to BBC show staff writer Anthony Coburn, who came up with the idea for the first episode after taking a walk and seeing a police box while on break from writing the show! The TARDIS, which in a 2011 episode written by author Neil Gaiman was revealed to have a sentient consciousness, was created by the Time Lords, the ET civilization Doctor Who is a member of, and its interior is vastly larger than its exterior. It takes the strange shape of a police box because of some faulty circuitry that trapped it in that particular form. The name TARDIS was trademarked by the BBC.

Figure 6-1 and Figure 6-2: An actual police box outside of Earl's Court, London, on the left, and a TARDIS used on the BBC series Dr. Who *on the right.*

On his adventures, the good doctor meets with many enemies, including the famed Daleks and Cybermen, in his ongoing attempts to make wrongs right and help those in need of a hero who can breach the boundaries of time. The original series, produced in the United Kingdom, ran from 1963 to 1989, was launched again via the BBC in 2005, and has spawned many spin-offs, including *Torchwood*, *The Sarah Jane Adventures*, and *Doctor Who Confidential*, among others. Interestingly, the very first episode of the show occurred the day after the JFK assassination, and the BBC re-aired that episode the following weekend prior to the second episode. One has to wonder if this had anything to do with influencing blockbuster author Stephen King to think about writing his newest novel, the time-travel themed, massive tome, *11/22/63*.

Dr. Who's TARDIS has now become all the rage, as has the show itself, on social network sites like Facebook and Twitter, where people are encouraged to build their own blue police box time travel unit and submit photos! The show is just as popular as ever, especially at Comic Con and other sci-fi-themed conventions, and proves that time travel–based entertainment will always fascinate and enthrall audiences of all ages.

But we are not science fiction characters with a long-running television series, and the troubling limitations of light and those nasty time travel paradoxes are like massive brick walls we cannot find a way over, around, or through.

Or can we?

The most challenging limitation of course is this pesky problem with light and the speed of light.

Light Speed Away!

Our understanding of light is being challenged by cutting-edge scientific discoveries every year. In Chapter 3, we talked about recent experiments that bend, disrupt, and manipulate rays of light to create cloaking of space and time. Yet it may even be possible to create light from a vacuum. In November 2011, the science journal *Nature* reported that scientists at Chalmers University of Technology conducted an experiment that captured photons appearing and disappearing in a vacuum. Led by physicist Christopher Wilson, the team succeeded in getting photons to leave their virtual state and exist as measurable light by allowing the virtual photons to bounce off of a mirror that moves near the speed of light. During the experiment, the mirror transferred some of its kinetic energy to the virtual photons, allowing them to materialize. This is known as the "dynamical Casimir effect." It was actually predicted more 40 years ago by physicists, but never attempted successfully—until this experiment.

This stunning experiment also emphasized that empty space is not necessarily empty, and the constant appearance and disappearance of virtual particles in a vacuum, known as vacuum fluctuations, could have a connection with the elusive "dark energy" that is behind the accelerated expansion of our universe.

A Sound Argument

Einstein stated that nothing could travel faster than light. He was referring specifically to particles and even information. But can sound travel faster than light? *Sound?* Apparently yes, and seemingly this does not violate the laws of physics *or* make Einstein want to turn over in his grave.

In 2008, John Singleton, a scientist and Los Alamos National Observatory Fellow, created a device that manipulates radio waves by forcing a change of state so substantial that they become superluminal transmissions that are faster than light speed.

Their polarization synchrotron combines radio waves with a rapidly spinning magnetic field that then transmits much the same way a pulsar, or rapidly rotating neutron star, emits radio wave pulses.

Along with colleague Mario Perez, Singleton presented the work to the American Astronomical Society at its major annual conference, and the two scientists are working on a series of even more powerful devices that may one day have huge applications in the field of technology.

Light again made headlines in April 2011 when Professor Akira Furusawa and his team at the University of Tokyo's department of applied physics succeeded in teleporting light. The experiments involved a machine known as "the teleporter" and basically destroyed light in one location, and re-created it at another location, rendering light both there and not there or, in the spirit of the classic Schrodinger's cat experiment, both "dead" and "alive" at the same time. Researchers in Australia also successfully conducted the experiment, which transfers a complex set of quantum information from Point A to Point B and could revolutionize quantum computing. The idea for this has been around for at least a decade—once again proving that, if we wait long enough, even something right out of *Star Trek* can become reality.

Meanwhile, quantum physicists at the University of California at Santa Barbara, led by Andrew Cleland and John Martinis, designed a "quantum machine," as they call it, that might one day lead to proof of time travel and parallel universes. Their machine, a tiny, little teleporter barely visible to the naked eye, involves making a tiny metal paddle cool to its ground state, the lowest energy state permissible by the laws of quantum mechanics, and then raising its energy slowly by a single quantum to produce a purely quantum state of motion. They even were able to put the device in both states at once, so it vibrated both slowly and quickly at the same time, in another sort of Schrodinger's cat state of superposition. They posited that we can only see one of these potential states at once, and upon the act of observation, the state then splits into additional universes. Perhaps, there is a plethora of multiple or parallel universes all around us, but we cannot see them.

Wormholes could also be another possibility for teleportation, as physicist Max Tegmark suggested while attending a panel, in January 2008, at MIT to discuss the science behind the movie *Jumper* starring Hayden Christiansen, about a man who can teleport all over the world at will. Tegmark was asked about the science behind the science fiction and remarked that a wormhole was one possible way of getting something quickly across space-time. However, after admitting that wormholes do appear to be theoretically possible, Tegmark commented that the actual trip would be rather grueling because of the instability of the wormhole: "It could collapse into a black hole, which would be kind of a bummer."

We like the whole parallel universe/Multiverse idea a lot better!

Not So Uniform

Before we tackle the Multiverse, it's critical to note that our own universe may not be as uniform as first thought, with laws of physics that are different in one end from those in another. Indeed, in July 2010, a team of Australian and British astrophysicists proposed just that when they revealed their evidence that the laws are different depending on where you are in the cosmic scheme of things. The three universities involved, thr University of New South Wales, Swinburne University of Technology, and the University of Cambridge, submitted their findings to the journal *Physical Review Letters*, stating that one of the finely tuned constants of Nature might not be so constant after all.

They focused on the fine-structure constant of "alpha," which appears to vary throughout the universe. This mathematical constant refers to the strength of electromagnetism, and they discovered that it varies continuously "along a preferred axis throughout the universe," as team researcher Professor John Webb reported. They utilized measurements taken at the VLT (Very Large Telescope) in Chile and Keck Observatory in Hawaii, and are awaiting peer review as of this writing, but if their findings hold water, we may be looking at new laws of physics for one location in the universe and the varying possibilities those new laws allow for. Webb posits that the implications are profound, not only because of new laws that may arise from this discovery, but also the idea that these varying laws may preclude the formation of life as we know it, and maybe even far, far different from what we can imagine.

In the quantum world, things behave in a spooky fashion, and quantum entanglement has proven that signals can move between particles simultaneously, and thus faster than the speed of light. But attempts to find this same spooky action at a distance as Einstein called it, and that some critics claim happens outside the usual space-time and therefore do not breach light speed laws in this space-time (huh?), in the larger scale world have tempted and tantalized many a scientists. The year 2011 proved to be a stunning advance forward when one particular and peculiar particle managed to do what no other had done before it.

But first, let's go back in time a few months earlier. Recall that in July 2011, Hong Kong physicists at the University of Science and Technology, let by Professor Du Shengwang, reported that they had proven that a single photon, the fundamental quanta of light, cannot travel faster than light—no way, no how. They used their findings to prove that time travel was impossible, and the science fiction world mourned.

Now we jump via wormhole to September of that same year, when an Italian experiment called OPERA (Oscillation Project with Emulsion t-Racking Apparatus) turned the science world upside down by announcing that a particle called a neutrino appeared to have done the unthinkable: surpass light speed. OPERA, an underground facility in the Gran Sasso National Laboratory in Italy, is a complex detection system of electronics and photographic emulsion plates that is located 454 miles from the CERN laboratory in Switzerland, which houses the Large Hadron Collider that in 2012 got a hint of the elusive Higgs Boson. But Higgs isn't the star here.

The Neutrino Experiments

The stars are neutrinos, electrically neutral fundamental particles that have no mass and yet are everywhere around us. The sun's nuclear reactions alone create billions of neutrinos that pass through your eyeball every second! They pass right through the Earth as if the planet were a vacuum. Neutrinos are everywhere. The OPERA got a blast of neutrinos from CERN and found, surprisingly, that the mysterious particles exceeded the speed of light, even if only by a teeny, tiny bit. We are talking 15,000 neutrino particles, which achieved a velocity of 20 parts per million faster than light speed. This comes out to 60 nanoseconds faster than the speed of light allows and marks the first time a particle has been able to move faster than light in a vacuum. There have, in the past, been experiments involving particles traveling faster than light through a specific medium—say, water.

Antonio Eraditalo, a physicist at the University of Bern in Switzerland and spokesman for OPERA, stated to various press outlets, including *Scientific American*, which covered the event in "Particles Found to Travel Faster than Speed of Light," that the research team was confident of their findings and had the same results in more than 16,000 events measured over 2010 and 2011. This parallels earlier findings in 2007 at the MINOS (Main Injector Neutrino Oscillation Search) experiment in Minnesota, which was sent neutrinos from the nearby particle physics facility Fermilab, in Illinois, and saw similar results. The neutrinos also arrived ahead of schedule, and at the time the MINOS research team did not play up the results because of some uncertainties with the positioning of the detector.

But MINOS does intend to follow up their experiments, and their hopes now may confirm the findings at OPERA, lay to rest the neutrino question, and open the door to a complete rewrite of the laws of physics. One theoretical physicist, Antonio Zichichi from the University of Bologna in Italy, told *Scientific American* that these "superluminal" neutrinos could be slipping through extra dimensions in space, which is part of string theory. And space scientist/author Dr. David Whitehouse stated that if the experiment stood the test of peer scrutiny and duplication, it would be "an earthquake, a revolution in physics. As soon as you think you are arrogant enough to think you understand the universe, the universe comes along and shows you are not right." Whitehouse was interviewed by Sky News for its online site (*news.sky.com*) on December 10, 2011, and even spoke about time travel as a possibility. "Everything is now open—because time, speed, and the speed of light are all linked."

Detractors and skeptics jumped on the announcement, with criticisms of inaccurate measurements, miscalculations of speed, and everything else, including revelations that came out early in 2012 about faulty connections of the fiber optics cable that brings synchronization to the GPS system used in the experiment, or possibly flaws with the oscillator within the master clock that may have given a false speed impression, but OPERA carried out further, improved experiments that adjusted the way the beams were produced, and the results were the same each of the 20 times the experiment was run. This ruled out some of the potential systematic errors that may have affected the results and gave even more credence to the findings.

Still, the world waits as other researchers at other sites attempt to duplicate the efforts, as planned by the United States at MINOS and the T2K in Japan, a particle physics experiment that is a collaboration between several countries, including Canada, France, and Germany. Take away any faulty cables or oscillators and—who knows—we may jump light speed yet!

■■■■■

The Large Hadron Collider is currently made up of six underground detectors. Two of them, the ATLAS and CMS (Compact Muon Solenoid), are large, general-purpose particle detectors. ATLAS will be utilized to look for the origins of mass and detect extra spatial dimensions as predicted by string theory. CMS will engage in the hunt for the elusive Higgs Boson or "God Particle" that recently took the media by storm, when hints of its existence stunned scientists, and will also look for clues into the nature of dark matter. Two are more specifically designed for studying what happened after shortly after the Big Bang.

ALICE will focus on the "fluid" form of matter called quark-gluon plasma that existed immediately following the Big Bang, and LHCb will look for what happened to the missing antimatter that was once thought to be in equal amounts with matter shortly after the Big Bang. The final two detectors, TOTEM and LHCf, are much smaller and are very highly specialized.

If in the end, after numerous experiments, the neutrino stands up as an FTL particle, how would it change the possibility of time travel from no to maybe-yes? Neutrinos are considered ghostly particles— neutral, without mass, and anomalous by their very nature of being able to change their type, or flavor, at will—but should they be proven beyond a shadow of a doubt to break the speed of light, the great cosmic constant, then everything we know about physics and our universe will change. That is nothing to scoff at. String theory, for one, will be a given, if it can be shown that these neutrinos are hopping into extra dimensions to take shortcuts to their final arrival point. Perhaps *only* neutrinos can do this, which would still be a scientific breakthrough of tremendous importance, but a letdown to those hoping other particles can do likewise.

And it's not like we could make a time travel machine out of neutrinos anyway. By their very nature they have no mass. A machine, of course, would. The possibility that a constant, or law, such as the speed of light might not be so constant after all, is enough to keep the research going and the theories coming, or vice versa if you prefer that order. Cause before effect...

Tachyons Revisited

Remember tachyons back in the 1960s? Those bizarre particles were, at the time, thought to be superluminal, but were never proven to even exist. In his mind-bending book, *Time Loops and Space Twists*, physicist Fred Alan Wolf, also known as "Dr. Quantum," discusses tachyons, particles moving faster than light, and that the special theory of relativity does not restrict them from being faster than light. But, tachyons would be subject to restrictions such as not ever being able to slow down to actual light speed because of the sheer amount of energy that would require, and they would possess imaginary rest masses; because they are never at rest, we would never be able to observe them as such! Tachyons become "virtual particles," and their processes are "virtual" processes as opposed to real processes.

Wolf also discusses the bizarre effect of combining quantum theory with relativity. Relativity says nothing can go faster than the speed of light—no particle or wave or physical process. But when you combine relativity with quantum physics, particles will have both positive and negative energy. "Since we've never observed particles with negative energy and we're not sure what that means, we can just assume that particles can't have negative energy and see what emerges from our theory with this restriction." But what does emerge, Wolf continues, are such things as quantum field theory, antimatter, and particles that can go backward in time. All of these have been experimentally verified as well, and thus, "If you only allow particles moving forward in time with positive energy in your theory, then you must also allow particles to go faster than light. But faster-than-light particles are really weird: they're called tachyons and have very strange properties."

Not everything has to travel slower than light; some particles can go faster than light, Wolf says, and it appears as if they are traveling backward in time. This particle may appear to us as going forward in a "unique time direction as an anti-particle going forward in time but with the opposite charge." Thusly, anti-matter may consist of particles of the same mass and charge going backward in time. In a wonderful statement, Wolf also admits that physics is a set of rules physicists come up with "by recognizing limits on the way nature has to behave." But, when confronted with said restrictions, often instead of making things more calm and simple, the result is more "seemingly impossible and counterintuitive!"

And what could be more seemingly impossible and counterintuitive than the idea that the past is not set in stone? Even freakier, the future may be influencing the past! Now we turn to photons, particles of light, which have been involved in some amazing experiments that truly twist our concept of the past creating the future.

In 2002 and again in 2007, scientists carried out numerous experiments involving particles of light that knew in advance what the distant twins would do in a future state. Researchers in France shot photons into an apparatus and were able to demonstrate that their behavior could retroactively change something that had already occurred. It worked like this: The photons passed a fork where they had to decide whether to behave like particles or behave like waves when the hit a beam splitter. Long after they made their decision and passed the fork, the experimenter could then randomly switch on a second beam splitter. The observer, or experimenter, it was discovered, decided at that point, when the beam splitter was switched back on, what the particle did at the fork in the past, and even *change the particle's behavior* even though it had already happened earlier in the experiment. Of course, these were all based upon the earlier double slit of Princeton University's John Wheeler, who coined the word *black hole*, if you recall. Wheeler's experiments set the stage for the theory that what happens in a quantum particles future will also affect its past. In the quantum world, the arrow of time does not exist, and, as Wheeler proved, the outcome of a particle experiment can be affected by a measurement that takes place even after the actual experiment did.

Wheeler often likened this to our observation of light from a distant object far out in the universe, and said that when we observe that light we literally make a quantum observation on a cosmic scale, and the measurements we make on this light today are actually helping to determine the path the light took billions of years ago. As we observe, we collapse the wave function and make a measurement, but a different person might collapse a whole different wave function, thereby creating a completely different measurement. We may be doing this with reality and history. Someone examining the death of a famous figure may, in his research, collapse only certain things yet not others into a fixed reality, yet he may not possess every possible collapsible state. Therefore, he may actually be shaping the past with his research by the very choices he makes as to what to investigate and what to leave alone!

Parallel Possibilities

Many scientists look to the possible existence of other levels of reality, or other universes, as a way to make time travel work outside of the restrictions of light speed and paradoxes. Imagine another universe alongside our own where the laws of physics are so completely different that what is impossible here is mundane and trivial there—multiple worlds, even, where each is different from the other, or perhaps an infinite number of universes where many would be exactly like our own. Hey, you might even exist in some of them just the way you are right now. In others, you might be rich, famous, handsome, or even a cockroach! In fact, perhaps you might even be invisible in one of them!

We are getting away with ourselves here, and when the talk turns to the Multiverse and other similar concepts, it's easy to start dreaming of science fiction worlds with every possible kind of life and all sorts of amazing machines and devices—and time travelers passing effortlessly back and forth between the past, present, and future as if it were nothing more than a visit to a few Saturday morning garage sales.

Because these terms are often used interchangeably to describe their commonalities, despite having some rather distinct differences, it's helpful to get a more detailed look at each.

Parallel Universes

As research fires up and continues at the Large Hadron Collider at CERN, near Geneva, Switzerland, one of the things physicists will be hoping for a glimpse of is a parallel universe, possibly hidden within other dimensions that go beyond length, depth, breadth, and time. The smashing of particles could open up a whole new world of understanding where they to reveal the existence of other universes if detectors were to see particles vanishing into these other possible dimensions and then returning back to the four we live with.

Parallel universes have long been a mainstay of science fiction films and stories. Parallel universes can exist individually, or grouped together as the Multiverse, and offer the possibility of a totally different reality in which someone, or something, can exist, or hop back and forth between. The laws of nature may be different in one parallel universe from what they are in another, and, with respect to time travel, would provide multiple versions of the future in which someone could exist, or not exist at all. Light speed limitations may not exist in a parallel universe, and the paradoxes that keep us from traveling back in time would be null and void if we could jump into a different historical time line.

Two great fictional examples of a parallel universe would be *Alice's Adventures in Wonderland*, written by English author Charles Lutwidge

Dodgson under the pseudonym Lewis Carroll, and C.S. Lewis's *The Chronicles of Narnia*, both of which involve some sort of portal or wormhole, such as a rabbit hole or a large piece of furniture, through which a person can enter into another realm.

Theoretically, parallel universes may be the result of a single random quantum event that branches off into an alternative universe. This is the "Many Worlds Interpretation," or MWI, and posits that each time a different choice is made at the quantum scale, a universe arises to accommodate that choice, thus creating infinite new worlds popping up all the time. These new worlds are being constantly created and could cause problems for a potential time traveler. Physicist David Deutsch wrote in "Quantum Mechanics Near Closed Timelike Curves" for the 1991 *Physical Review* that, if time travel to the past were indeed possible, the many worlds scenario would result in a time traveler ending up in a different branch of history than the one she departed from. Deutsch, of Oxford University, is a highly respected proponent of quantum theory and suggests quantum theory does not forbid time travel, but rather sidesteps it, referring to the traveler's ability to go into another universe—a parallel universe—and avoid the paradox limitations. Deutsch has shown mathematically that the branch-like structure of one universe splitting into another parallel version of itself may also explain the probabilistic nature of quantum outcomes, which has been confirmed by the research of fellow Oxford physicists David Wallace and Simon Saunders. All of this has given more credence to the MWI.

Many, Many Worlds

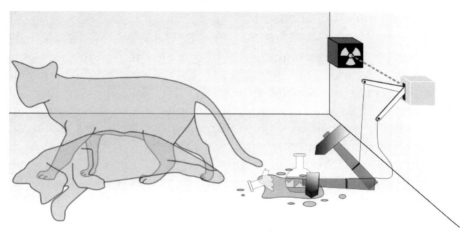

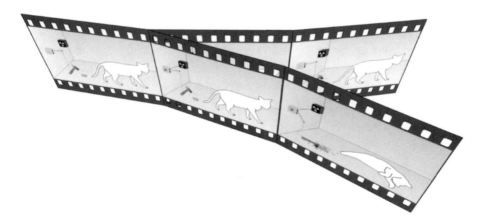

Figure 6-3 (top) and Figure 6-4 (bottom): The famous Schrodinger's Cat experiment shows how the multiverse works by splitting off into two worlds, one where the cat is alive in the box, and one where it is dead.

In 1957, physicist Hugh Everett III proposed a quantum interpretation as part of his Princeton University doctoral dissertation that suggested that reality might be more like a tree with many, many branches representing alternative universes each with their own reality and history and time line, not like a flagpole with only one. Everett's theory looked at the famous Schrodinger's cat experiment and suggested that every event is a branch point for a new universe to form—one in which the cat is dead, one in which the cat is alive, by having both the dead and alive cats exist at the same time—until the act of observation opens the box and sees the reality. The cat was either alive or dead. But the many worlds theory took away the observer-driven choice and denied this "wave function collapse" by which the cat is "measured" as dead or alive. In this theory, it is both and everything else, all at the same time, but the alive and dead cats exist in two separate universes.

It was actually Bryce Seligman DeWitt who named this theory in the 1960s, but Everett popularized the concept of branching worlds that may be infinite in number and contain every possible event that could ever happen, and even contain us in a few of them. We may have more than one doppelgänger!

This theory operates at the quantum level, and yet looks at the universe as being quantum mechanical in nature. Everett focused on the universal wave function, which is the mathematical list of every possible configuration of a quantum object, such as an elementary particle, and allows for everything that is possible to be possible. Take an electron, for instance. It could exist simultaneously in different orbits, just as Schrodinger's kitty cat could exist simultaneously as both alive and dead. Everett, and those who support and have built upon his theory, wanted to tackle the problem of wave function collapse and superposition, and why an observer would only see one outcome or reality of a particle's position once the function was collapsed. Each time the observer looked at the particle, he or she might see a different outcome or reality. Instead, he proposed that—at the quantum level, at least—there is nothing but a huge superposition of every particle in the universe and every single possible position those particles could have, which is mind-boggling. When the wave function collapses it chooses one state upon measurement. But in the MWI, the initial mix of these states never collapses, and instead, when a measurement is made, it splits our universes into parallel universes in an abstract quantum level and *all* possible outcomes occur somewhere in that quantum level albeit on a different universe.

In terms of time travel, we might extrapolate that these new-branched universes are realities we can exist in on a human scale, and that each has its own time line, history, and order of events different from the next. Infinite time lines in fact....

Deutsch's idea of parallel universes, the Multiverse, or "shadow universes" was described in his interview with the *Guardian UK* in June 2010 ("David Deutsch's Multiverse Carries Us Beyond the Realm of Imagination") as being "co-incident with, somehow contiguous with, and weakly interacting with, this one. It is a composite, a layer cake, a palimpsest of universes very similar but not quite identical to each other." The number of these shadow universes could be enormous, and Deutsch points to photon experiments that suggest possibly a trillion of them or more. He also suggests that future-directed time travel will essentially only require efficient rockets, and is on the "moderately distant but confidently foreseeable technological horizon." When it comes to past travel, the Multiverse might save a time traveler from the pesky grandfather paradox. He uses an example of a writer who wants to go back in time with a copy of Shakespeare's Complete Works and help the bard complete *Hamlet*. It can happen, but in the Multiverse view, "the traveler has not come from the future of that copy of Shakespeare."

Another offshoot of the MWI is the Many-Minds Interpretation, which extends the MWI by positing that the branching off of worlds occurs in the mind of the individual observer, introduced in 1995 by theoretical physicist H. Dieter Zeh, professor emeritus of the University of Heidelberg and the discoverer of decoherence. The Many-Minds Interpretation was widely criticized and somewhat ignored, mainly because of issues involving the theory that the mind can supervene on the physical as the mind has its own "trans-temporal identity." The mind may select one identity as its own non-random "reality," yet the universe as a whole remains unaffected, which presents additional problems when dealing with different observers ending up with the same measured realities. The actual process by which the mind of the observer would select the single, measured state is not explained by the MMI.

M Theory

One of the pet theories in theoretical physics that allows for parallel universes to exist is M Theory, or Membrane Theory (although the "M" often stands for Mother, Magic, and Mystery!), which posits that there are many universes that take the form of three-dimensional branes and exist alongside each other in another "fourth" dimension. M Theory is an extension of string theory and was first suggested in 1995 by physicist Edward Witten of the Institute for Advanced Study. Witten drew on the earlier research of such luminaries as Michael Duff, Paul Townsend, Ashoke Sen, John Schwarz, and Chris Hull, all of whom had ventured into string theory territory and helped launch the "superstring revolution." Each membrane might have its own natural laws and even history, so again the question would be: How can you get from one brane to another? Some physicists believe that each time two branes collide, a new Big Bang occurs, thus creating a new universe, and that our own Big Bang may have been a brane collision! M Theory also posits 11 dimensions (up until M Theory was introduced, string theory suggested 10 dimensions were necessary) and unites five prior and conflicting string theories into a unified "superstring theory."

Alternate time lines, each with its own forward arrow of time and its own history, may exist then, allowing time travelers to jump into another version of history and override those pesky paradoxes. Imagine being able to jump into a time line where you get your dream of marrying your high school sweetheart, but finding out she's an evil tramp as soon as you say "I do." You could jump back into your original historical time line, where you didn't marry her and instead ended up three years later marrying her sister, your true soul mate, and lived happily ever after.

It could happen!

Paths Around Paradoxes

Noted theoretical physicist Michio Kaku, author of *Parallel Worlds*, *Beyond Einstein*, and *Hyperspace*, win his newest book, *Physics of the Impossible: A Scientific Exploration Into the World of Phasers, Force Fields, Teleportation and Time Travel*, offers three ways around the paradoxes of time travel. The first is that you simply repeat past history and fulfill the past, and that everything you do once you are back in time was meant to happen anyway, a sort of destiny.

This opinion is also mirrored in the views of famed physicist and superstring theory proponent Brian Greene, author of *The Fabric of the Cosmos: Space, Time and the Texture of Reality* and *The Elegant Universe: Superstrings, Hidden Dimensions and the Quest for the Ultimate Theory*. Greene writes that outside of the quantum world, in the classical science of the grander scale, we exist—static and unchanging—at various locations in what he calls the "space-time loaf" of block we call space-time. These moments are unchangeable and fixed. Using a wormhole, if one were to indeed go back in time to a certain point, or date, one would find that there is only one version of that date and that your presence back in time would simply be a part of the original version of that moment. That moment has one incarnation, though. In *The Fabric of the Cosmos*, Greene writes, "By passing through the wormhole today and going back to that earlier time you would be fulfilling your iron-clad destiny to appear at that earlier moment." He points to the wormhole time machine itself as the culprit, with one opening or the other passing through time more slowly than the other end, but each opening still going forward in time. Thus, there will also be a limit as to how far back in time you could travel in the first place.

The second of Kaku's paths around the paradoxes involves having some free will to change the past, but within limits—so that you could go back and try to kill your grandfather, but something would prevent you from doing so. The gun might lock up, or you might drop it and shoot your foot instead and end up in the hospital. No matter what, you would somehow be prevented from knocking off your grandpappy.

The third involves the universe splitting into two universes to accommodate the time traveler. His example offers someone going back in time to kill their parents, and in one time line the people look like your parents but are different because you exist in a different timeline.

Kaku also suggests that the many worlds approach could solve all the paradoxes in two ways. First, if we imagine the time line of our universe as a line drawn on a board, then we can draw another line to represent the universe that branches off from the first. "Thus, whenever we go back into the river of time, the river forks into two rivers, and one time line becomes two time lines, or what is called the 'many worlds' approach." Say you planned to kill your own father. You go back in time, and you do the dirty deed. Kaku states that if the river of time does indeed have many forks, this would not be a problem. "You've just killed somebody else's father. In that timeline, you don't exist, but you exist because you jumped the stream."

This idea would also, Kaku proposes, solve another thorn in the side of physicists when discussing time travel: the radiation effects of entering a wormhole, which would no doubt destroy any time traveler, and also end up in a loop, the feedback of which would become so strong it would destroy the wormhole. "If the radiation goes into the time machine, and is sent into the past, it then enters a new universe; it cannot reenter the time machine again, and again, and again." Kaku points out that the main problems involving time travel and wormholes specifically center on the issues of the physics of the event horizon, as in the stability of the wormhole, the deadly radiation, and the wormhole closing once it was entered. Solve those issues and really time travel might be a piece of cake!

Well, not a piece of cake, but all physicists agree that once they come up with a Theory of Everything that unites the four universal forces of electromagnetism, gravity, and the strong and weak nuclear forces, and formulate a complete theory of gravity and space-time, then time travel might be as close as finding a wormhole big enough, stable enough, and open enough to get a time machine through. Not to mention the sheer amount of energy required! This might require harnessing the power and energy of a neutron star, or finding that elusive exotic matter scientists are looking for or a good source of negative energy, and we are far from doing so.

Then there is the problem of creating the machine, and let's not forget finding or creating a wormhole that could handle it! An interesting problem was brought up by physicist and cosmologist Paul Davies, author of *About Time: Einstein's Unfinished Revolution* and other books. In an interview with Discovery.com called "Is Time Travel Possible?" he discussed wormholes as time machines and potential time travel tourists from the future, but with the caveat that "theoretically, it would take more than 100 years to create a 100-year time difference between the two ends of a wormhole, so there's no way that our descendants could come back and tell us we're wrong about this." So, it's all about timing, then (pun intended).

When it comes to parallel universes, Kaku points to the three types addressed in scientific arenas as:

1. Hyperspace/higher dimensions.
2. The Multiverse.
3. Quantum parallel universes (many worlds).

Do higher dimensions exist? If we buy into superstring theory, they would have to—at least 10 of them to make the theory mathematically workable. These dimensions, which physicists at CERN's Large Hadron Collider also hope to see a glimpse of, could include not only extra spatial dimensions, but perhaps an extra dimension of time as well. Most agree that these dimensions might be anywhere from so tiny and curled up we would never be able to see or access them, to infinite in size and existing right at the tips of our own noses.

On a Mission of Time

One of the most influential figures in time travel research is a man on a mission—a personal mission to try and make time travel a reality, maybe even in the 21st century. Dr. Ronald Mallett, a physics professor at the University of Connecticut, has authored a book, with Bruce Henderson, titled *Time Traveler: A Scientist's Personal Mission to Make Time Travel a Reality.* In the book, Mallett documents his search through the bizarre and intriguing worlds of wormholes, cosmic strings, general relativity, quantum gravity, and black holes in an attempt to design a time machine with circulating laser beams, and test it sometime by the year 2016 or so.

He has more than just a passion for his work, though. He has a personal reason that drives his quest to master the challenges, obstacles, and paradoxes standing in the way of traveling through time. When Mallett was 10 years old, his father, Boyd, an electronic technician, passed away of a heart attack at the young age of 33. Boyd was little Ron Mallett's world, and Mallett was inspired by a comic book version of the classic H.G. Wells's *The Time Machine* to find a way to go back in time, help his father live a healthier lifestyle, and save him from that fatal heart attack. In one of the opening chapters of his book, he tells the touching story of how his father conveyed to him the importance of education by pointing out to little Ron a new highway construction project outside the living room window. Boyd pointed to a team of ditch diggers and asked his son, "Is this what you want to do?" Ron told him no, and his dad responded, "Then you better take multiplication and school more seriously." Ironically, Ron hated math and didn't even think about going into physics during his high school years. But his desire, upon his father's death, turned into a magnificent obsession that led to a PhD from Penn State University in 1973, and his full professorship at the University of Connecticut. He is also a member of the American Physical Society and the National Society of Black Physicists.

Mallett's book is a detailed examination of the research and theories that propel him toward the day when he might be able to utilize what we already know about Einstein's general relativity, upon which he bases much of his work, coupled with exciting and innovative ideas involving circulating lasers that, at sufficient energy levels, might produce closed timelike curves (CTCs), which we wrote about in previous chapters. Rick Steinick of Decoded Science interviewed Mallett in October 2011 about the progress of his experiment, involving researchers at both Penn State and the University of Connecticut and his theory for making his dream come true. Mallett said: "The concept using a circulating beam of laser light to twist space and time is straightforward. However, the theoretical details are quite involved and the experimental implementation of the concept is challenging." He went on to say that the machine designed for the experiments uses light in the form of circulating lasers to "warp or loop time," and acts as the third piece of a time travel puzzle that also involves Newton's law of universal gravitation, and Einstein's rule of general relativity in a chronological time line.

Mallett states that his time travel theory is a sort of culmination of our understanding of gravity first proposed by Newton, and then more deeply understood by Einstein's work. He also suggests wormholes and potential parallel universes as viable shortcuts if we could find them traversable for a human time traveler. He teaches his students the importance of realizing that making time travel a reality rests upon solid physics.

He calls his project STL, for "The Space-time Twisting by Light," and progress is ongoing as the necessary funding is acquired. By the time you read this book, perhaps Mallett will have already begun his trip back in time to the year he lost his beloved father, and test out those pesky paradoxes in person.

The Multiverse

The Multiverse is the most widely mentioned theoretical "time travel paradox killer," because it involves more than just one parallel universe, thus allowing for an increasingly possible world where the laws of physics are just right for time travel—if we can get from here to there.

There may be a massive number of other universes out there, possibly even an infinite number, or maybe just 20 or 70. While our astronomical observations cannot at this time detect them, it is most definitely a theoretical possibility that many cosmologists and physicists are considering. These universes may or may not be like ours. In fact, they may or may not even have the same laws of physics or distribution of matter, or even number of spatial and temporal dimensions. Some will undoubtedly be "dead" and others will have life forms that we cannot recognize or even imagine. Others still may have duplicates of us living their own separate lives and time lines. Maybe Big Bangs are going on constantly, 24/7/365, all the while creating new universes. Just imagine how many new worlds are created while we brush our teeth, make our salami sandwiches, and watch the Super Bowl.

Imagine a bunch of "you's" reading this book all at the very same time, and a host of other "you's" doing a host of other really cool things in worlds beyond our perception and detection—at least scientifically. In Chapter 8 we'll explore some more "paranormal" ideas about perceiving other realities and time lines!

The Multiverse theory is not new, especially for readers of science fiction and fantasy, where other worlds beyond ours is a given. The actual term was coined in the year 1895 by psychologist and philosopher William James, and is now a mainstay of theoretical and quantum physics, as well as a part of our religious beliefs, mythological stories, and spiritual/New Age thought. The Multiverse has been equated with everything from the Kingdom of Heaven of the Judeo-Christian Bible to the various planes of existence of more metaphysical and spiritual thought to the multiple time lines and dimensions of more paranormal and anomalous concepts.

Cosmologist Max Tegmark took the Multiverse theory to the next level by creating a classification level for potential other worlds:

Level One: Domains beyond our cosmological horizon—the least controversial type, what lies beyond the vantage point, yet likely has the same laws/constants, just with possibly different initial conditions than our own.

Level Two: Universes with different physical laws/constants, other post inflation bubbles; far more diverse than Level Ones; these bubbles also vary in initial conditions as well as other seemingly immutable aspects of nature.

Level Three: Quantum universes/Many Worlds Interpretation exist alongside us on the quantum level where the random quantum processes cause the universe to branch into multiple copies, one copy for each possible outcome.

Level Four: Ultimate Ensemble—other mathematical structures, where *all* potential alternate realities can exist, anything and everything is possible in terms of location, cosmological properties, quantum states, and physical laws and constants. These exist outside of space-time.

Each level of Multiverse has its own characteristics that separate it from the other levels, and for our purposes the focus for time travel would be on those we humans could exist in, and possible travel between. One of the ways Tegmark himself differentiated the levels was by stating that in Level One, our doppelgängers could live somewhere else in three-dimensional space, but in Level Three they would live on another quantum branch in an infinite-dimensional Hilbert space, yet, as the Many Worlds Interpretation states, likely not be able to interact once the split into another branch occurs. Those found in Level Two might be like bubble universes that have different physical laws and constants, and each new bubble is created by splits that occur when spontaneous symmetry breaks occur in Level Three.

Tegmark describes these levels in detail in his book *Universe or Multiverse*, and states that the key question isn't so much whether there is a Multiverse, but rather how many levels it has. The book is full of deep physics, but one opening statement truly sets the stage for all theoretical inquiry: "Hubris and lack of imagination have repeatedly caused us humans to underestimate the vastness of the physical world, and dismissing things merely because we cannot observe them from our vantage point is reminiscent of the ostrich with its head in the sand." Tegmark admits that nature may have tricked us into thinking that our vantage point was the extent of reality, a fixed view of the world around us. "Einstein taught us that space is not merely a boring static void, but a dynamic entity that can stretch (the expanding universe), vibrate (gravitational waves), and curve (gravity)."

Life, and maybe even us to be exact, would exist on those universes that are fine-tuned for the existence of stars, planets, and biological life to exist, the same way our own universe seems fine-tuned. Paul Davies wrote about this extensively in his book *The Goldilocks Enigma*. Davies also wrote in the April 2003 *New York Times* article "A Brief History of the Multiverse" that science is hard-pressed to prove the existence of other universes, especially an infinite number of them. "The multiverse theory may be dressed up in scientific language, but in essence it requires the same leap of faith," he writes, referring to more theological discussions of the Multiverse.

In August 2011, *Science Daily* reported on the publication of two research papers in the *Physical Review Letters* and *Physical Review D*, which were considered the first to detail how a search for the signatures of other universes might be undertaken. A team of cosmologists from the University College London, Imperial College London, and the Perimeter Institute for Theoretical Physics looked at the disk-like patterns of cosmic microwave background radiation (CMB) that could be considered evidence of collisions between our universe and others. A collision would appear as a temperature modulation in the CMB temperature map used in the research, and must be distinguished from random patterns in overall data of CMB radiation across the sky.

The team ran a number of simulations of what the sky would appear like with and without these "cosmic collisions," and the result was a groundbreaking algorithm that was used to place an upper limit on how many bubble collision signatures there could be. The computer algorithm searched for these signs of bubble universe collisions, and although the team stated their research was far from conclusive, they were thrilled with the newfound ability to detect signature data, which could then be corroborated with incoming data from the European Space Agency's Planck satellite.

Many scientists refer to the Multiverse as more of a "pocket universe" concept, indicating different regions in space-time that are unobservable, but still a part of our one universe. Inflationary cosmology does state that these pocket universes can be self-contained, with different laws of physics, different particles and forces, and possibly even different dimensions.

Even the popular string theory allows for potentially trillions of possible universes, each one compatible with relativity and quantum theory. Michio Kaku states in *Physics of the Impossible* that "Normally communication between these universes is impossible. The atoms of our body are like flies trapped on flypaper. We can move freely about in three dimensions along our membrane universe, but we cannot leap off the universe into hyperspace, because we are glued onto our universe." Gravity, however, can freely float into the spaces between universes. Kaku also points to one theory where dark matter, which is an invisible form of matter surrounding our galaxy, might actually be "normal" matter in another universe.

Multiverse and Entropy

Back to the question of time and time travel, though, and we once again return to entropy—the arrow of time and increasing entropy, to be exact, which we find to be the law of the land in our Universe. But what about the Multiverse? If there truly are an infinite number of universes, bubble or otherwise, would some of them be so different from our own that they would *note* start out in a low entropy state? Sean Carroll, referred to earlier in this book, is looking into the possibility of a model of the Multiverse that contains a "parent universe," one with no overall arrow of time. Carroll told Miriam Frankel of FQXi.com in a June 17, 2010, article titled "Time and the Universe" that if you start with a space-time with no directionality of time, then new baby universes would preferentially begin in a state of low entropy and, once grown, birth new universes as a result of the quantum fluctuations in the presence of vacuum energy. "These universes have arrows of time pointing to different directions, so, in some universes, time could actually run backwards." In the very same article, physicist Laura Mersini-Houghton, of the University of North Carolina, Chapel Hill, discussed her own research, funded by FQXi and involving the arrow of time. She posits that in the larger Multiverse, time is arrowless, favoring no particular direction. In her model, high-energy bubble universes overcome the "matter crunch" between expansion and inflation, and matter causing the bubble to try to "crunch back down." Only high-energy bubble universes become full-blown universes, again with low entropy as their starting state. But in the larger Multiverse, only baby universes retain the time-symmetry. The bubble itself does not, losing information about the universe from the moment of birth, leading to growing disorder, thus creating a local arrow of time as entropy increases.

Although his might seem difficult to comprehend, the basic idea is that perhaps there are ways around the low to high entropy state that we know and love so well as our universe. According to the concept, if this is the case, then time itself can run in any which way it pleases.

But the question remains: Can we travel back and forth between these different worlds with different laws and arrows of time? Again, theoretically, it would require a shortcut through space and time—like a wormhole—and a means of safely getting through that wormhole should it be stable and traversable. So even though the Multiverse theory takes care of some of the paradoxes by offering alternate time lines and histories in which one can both go back to the past and kill his or her grandfather (while not killing him at the same time), it appears as though there is still no realistic way of actually doing that.

The Multiverse also allows for alternate futures as well and for multiple, alternate versions of "you" to exist in any number of historical time lines with different outcomes, depending on the choices you make in each baby bubble universe. In an article titled "Riddles of the Multiverse" for PBS.org's August 2011 *NOVA* series, University of Southern California Professor of Physics and Astronomy Clifford

Johnson was asked straight out about whether or not the Multiverse could ever be "visited" by humans. His response was that we must first work out the physics of these other universes, in order to determine when and whether it makes sense to "cross over from one to the other." He did admit that it is possible that the stuff we are made of—the matter and forces that make us and hold us together—may not allow us to ever leave our four-dimensional universe and go to another. Imagine doing so and, well, coming undone! He points to gravity as something that might pervade all other universes. "So you can imagine scenarios in your wild dreams, where somehow we communicate with other universes through gravitational effects. But again, this is very speculative, and we're nowhere near where we need to be to make sense of this."

For now, it seems, we just don't yet have the brainpower and technology to leap and jump between worlds, to cross time lines and experience as many pasts, presents, and futures as we would like. As Kip Thorne stated in an October 1999 PBS.org NOVA "Time Travel" panel interview with colleague Carl Sagan, "There are several different ways to turn a wormhole into a time machine if you are a clever and infinitely advanced civilization. By an infinitely advanced civilization I mean, somebody who can do anything their heart desires except they can't violate fundamental laws."

And yet, there are those who claim they already have done so.

7

TIME IS ON MY SIDE

A conspiracy is nothing but a secret agreement of a number of men for the pursuance of policies which they dare not admit in public.
—Mark Twain

Secrets are made to be found out with time.
—Charles Sanford

Man can believe the impossible, but man can never believe the improbable.
—Oscar Wilde

Do you believe that our government or its representatives would willfully and purposely deceive the fine citizens of our country in an effort to hide its activities? Has our government ever lied to its citizens in the past? Unfortunately, the answer to this would be a resounding *yes*(!), and details could fill an entire book in itself. Think about some of the more public examples, such as Watergate, the Iran Contra affair, and even the televised press conference when Bill Clinton artfully proclaimed, "I did not have sexual relations with that woman." Whether under the cover of "national security" or other trumped-up dictum, our government has (and most likely will continue to) hidden and obfuscated facts in an effort to "protect" us.

Let's take that one step further. Has the government achieved a working time machine that has been operation for several decades? Has DARPA created a time machine, as one popular conspiracy theory suggests? Are there actual human guinea pigs walking the Earth that took part in these top-secret projects? And were those aliens who crashed at Roswell really humans—from the future? Some say yes, and, although there are little facts or proof to back up their claims, they are stunning claims that definitely deserve a deeper look.

Since the dawn of mankind, conspiracy theories have been an integral part of nearly every significant event in our history. Conspiracy theories have served to elevate the most mundane affair into full-blown incidents. As we shall see, this pop phenomenon has spilled over to time travel as well. We love a good conspiracy, knowing perhaps on some deeper level that where there is smoke, there is often fire—or at least the beginnings or endings of a fire that may have burned rather brightly, even if it did burn in an underground secret facility or in another part of the time-space continuum! Suspend your belief for a time and come with us on a wild ride into some of the more intriguing time travel conspiracies out there.

Disclaimer: We authors, and our publisher, in no way agree with or support these claims. Our job is to present them, and let you decide. Some of these may seem crazy—insane, even—but we feel it is our duty to present them regardless. Buckle up!

A Time Travel Circus

Recently, conspiracy theorists were atwitter all over the Net and on social networking sites discussing a viral video clip from *The Circus*, a Charlie Chaplin silent film from 1928. What caused all the fuss was a very brief scene that some believe shows a woman holding a cellular telephone to her ear while walking down the street. This video clip quickly became a hotbed of controversy as some stated the simple, and others insisted on the bizarre, as a way of explaining the woman from our cinematic past clearly using what appeared to be technology from the present. Unquestionably, the video does show an individual holding some type of device to her ear (which seems to favor the shape of a cell phone) and talking into it. Incredibly, this film is one of the highest grossing silent films of all time... yet no one had spotted it before! Because cell phone technology had obviously yet to be invented back in 1928, some believe this person absolutely *had* to have been a time traveler.

Figure 7-1: This image from an old Charlie Chaplin film caused a viral sensation over whether or not the woman in question was a time traveler.

If you have read our previous books (and if you have not, we will wait right here while you go out right now and order them!) you already know that we are big fans of Occam's Razor. Occam's Razor (also known in Latin as *lex parsominiae*) refers to the principle whereby the simplest explanation is generally (and most likely) the most correct one. In other words, why on earth choose the most outlandish explanation when there were so many more common-sense explanations to choose from?

In fact, there is a simpler explanation that very well may explain the Chaplin time traveler: Four years prior to the release of *The Circus*, Siemens, the multinational corporation responsible for designing the first long-distance telegraph system, filed a patent for "a compact, pocket-sized carbon microphone/amplifier device suitable for pocket instruments." On their Website, Siemens describes the device as follows:

For a while, the carbon amplifier patented by Siemens played a major role in hearing aid technology and significantly raised the volume of hearing aids.

The electrical energy controlled by the carbon microphone was not fed to the received directly. It first drove the diaphragm of an electromagnetic system connected to a carbon-granite chamber. Current was transmitted across this chamber from the vibrating diaphragm electrode to the fixed electrode plate.

The amplified current produced mechanical vibrations in the electromagnetic hearing diaphragm that were then transmitted to the ear as sound.

A-ha! So, was our mysterious subject using a portable hearing aid? By the end of 1929, Siemens had produced several different products designed to enhance one's hearing ability. Other companies were likewise developing products for the hearing-impaired. Western Electric released the "Model 34A Audiophone Carbon Hearing Aid" back in 1925, which measured approximately 8 inches by 4 inches, and could very easily have been held in one hand, as it weighed less than 2 pounds when equipped with batteries.

In the video, the woman who is utilizing the device appears to be quite well dressed (wearing a top hat and long fur coat). Would it be a stretch to imagine that a socialite of her status would have almost certainly been able to afford such a luxury?

Detractors of the hearing aid theory point to several compelling facts surrounding the incident. As we know, the film premiered on January 26, 1928 at Grauman's Chinese Theatre. Checking the weather archives for Los Angeles, one will find that the temperature ranged from 70 to 80 degrees Fahrenheit during the last week in January 1928. Considering the weather conditions, why would this individual be wearing such inappropriate clothing? Furthermore, not only does she seem to be engaged in an animated conversation, but, her gait, mannerisms, and stature seem very masculine. Was "she" really a "he" in disguise?

Even if we ignore the glaringly obvious technical issues (the biggest being the lack of the necessary infrastructure to support a cellular network in 1928), the size of the device itself seems to be quite large by today's standards. If this individual truly was a time traveler, and had access to advanced time travel technology, why would she use an antiquated device seemingly from the mid-1990s? Would she not instead be rocking a killer iPhone or Droid from the future? That is an excellent question, but not the first time that a supposed time traveler utilized ancient technology.

■■■■■

Anatomy of an Urban Legend

Once upon a time, a man from the 1800s vanished into thin air, only to reappear on a busy New York City street 74 years later. His name was Rudolf Fenz, and in the middle of June 1950, he turned up in New York's Times Square stunned and dazed, standing in the middle of an intersection. Witnesses say he appeared out of nowhere, and none had time to step in and save him from being hit and killed by a taxicab as he stood there, a man out of time in more ways than one.

The story continues that he was dressed in clothing of the late 19th century, and had on his body old bank notes, a copper token with the name of a saloon on it, a bill for the care of a horse and washing of a carriage, a letter from June 1876, and his business card, with his name and address on Fifth Avenue. After his death, Captain Hubert Rihm of the NYPD Missing Persons Unit tried to find out who the mysterious man was, but could find no record of his address or fingerprints, and no one had reported Fenz missing. Finally Rihm was able to contact the widow of the alleged man's son, and she confirmed that her husband's father had vanished into thin air in 1876 at the age of 29. Rihm checked the missing persons files for that year and found that the description matched his appearance, age, and clothing. Rihm never made his findings official, and the case went unsolved.

Time-shift to the year 2000, when folklore expert Chris Aubeck did a little investigating of his own and found that the entire event was fictional. He traced the Fenz story to some of its origins and representations all the way to a 1972 issue of the *Journal of Borderland Research Foundation*, which sourced its version of the story to a book published in 1953 called *A Voice from the Gallery*, by Ralph M. Holland. End of story.

Well, not quite. Time-jump to 2002, when Aubeck had published his findings in the *Akron Beacon Journal*. That's when another source pinpointed the exact origin of the Fenz legend as a science fiction short story from author Jack Finney titled "I'm Scared." End of story.

Hoaxes such as this, whether intentional or just the result of people perpetuating rumors, spread more quickly now with the Internet, YouTube, and viral videos. What is the moral of the story? Sometimes the easiest explanation is the correct one, even if it does take all the fun out of a good urban legend.

The John Titor Enigma

John Titor. The name may not ring a bell, but to the time travel conspiracy world, that name may go down in history as one of the most controversial of our time. Although there is considerable debate over the exact date that it began, an individual by the name of John Titor (who initially used the pseudoname Timetravel_0) began posting on a public Internet forum that he was a time traveler from our future. In fact, Titor claimed to be from the year 2036!

This is a copy of Titor's initial Internet posting, from *www.johntitor.com*, dated January 27, 2001 at 12:45 p.m.:

> "Greetings. I am a time traveler from the year 2036. I am on my way home after getting an IBM 5100 computer system from the year 1975.
>
> My "time" machine is a stationary mass, temporal displacement unit manufactured by General Electric. The unit is powered by two, top-spin, dual-positive singularities that produce a standard, off-set Tipler sinusoid.
>
> I will be happy to post pictures of the unit."

According to Titor, he was an American soldier from the year 2036 living in Florida and assigned to a special governmental time travel project. Titor claimed that he had been sent back to the year 1975 to retrieve an archaic IBM 5100 computer system, which scientists of his day required in order to "debug" various legacy computer programs that were being used in 2036. It certainly does baffle the mind to think that a civilization capable of developing advanced time travel technology would somehow be reliant upon such an antiquated device, however, we digress. Interestingly, Titor claimed that he had made a layover in the year 2000 for strictly "personal reasons."

Titor posted convincing photos of his time machine (which, incredibly enough, was supposedly installed in the rear of a 1967 Chevrolet Corvette convertible) as well as its instruction manual. When asked, Titor provided more details of the device, stating that it contained:

- Two magnetic housing units for the dual micro singularities.
- An electron injection manifold to alter mass and gravity micro singularities.
- A cooling and X-ray venting system.
- Gravity sensors, or a variable gravity lock.
- Four main cesium clocks.
- Three main computer units.

During the course of the next several months, Titor became increasingly more public by posting numerous predictions to many Websites and message forums. According to the John Titor Webpage on Squidoo.com, some of his predictions included the following:

- "A world war in 2015 killed nearly three billion people. The people that survived grew closer together. Life is centered on the family and then the community. I cannot imagine living even a few hundred miles away from my parents." (November 4, 2000)
- "... the ice caps are not melting any faster than they are now. There is also far less smog and industrial waste in 2036." (November 6, 2000)
- "Life is much more rural in the future but 'high' technology is used to communicate and travel. People raise a great deal of their own food and do more 'farm' work. Yes, compared to now, we do work long hours." (November 7, 2000)

- "There is a civil war in the United States that starts in 2005. That conflict flares up and down for 10 years. In 2015, Russia launches a nuclear strike against the major cities in the United States (which is the 'other side' of the civil war from my perspective), China and Europe. The United States counter attacks. The US cities are destroyed along with the AFE (American Federal Empire)...thus we (in the country) won. The European Union and China were also destroyed. Russia is now our largest trading partner and the Capitol of the US was moved to Omaha Nebraska." (November 7, 2000)

- "Far less medical treatment in the future even though It's [*sic*] more advanced. People die when they now [sic] it's time to die. No lasers. Genetic medicine and cloning organs are the obvious new techs in the future." (November 7, 2000)

- "The Constitution was changed after the war. We have 5 presidents that are voted in and out on different term periods. The vice president is the president of the senate and they are voted separately..... If I could bring some material thing back to your time from 2036, it would be a copy of the new US Constitution." (November 7, 2000)

Titor continued to post detailed information regarding a variety of topics, including the process of time travel itself and how it might be achieved. Again, according to the John Titor Squidoo page:

- "While the machine is on, everything is black. When the machine is turned off, it is the reverse affect. It appears you are driving out from a bridge. To tell you the truth, I'm usually sleeping when the unit turns off but yes, it does appear that the world fades in from black." (November 4, 2000)

- "The only real physical trace is a large chunk of ground missing from the point of origin and a large pile of dirt at the destination. The gravity field surrounds a small portion of the earth under you and takes it along for the ride. There is really no way around this." (November 4, 2000)

- "When traveling to other worldlines there is a system of clocks and gravity sensors in the machine that sample the environment before dropping out. It's called VGL, (variable gravity lock). If a cement block were there, the machine would 'backtrack' until it sensed relative congruity to the original gravity sample. A great deal of time and effort goes into picking just the right spot since you cannot physically move during a displacement." (November 6, 2000)

- "Time travel is achieved by altering gravity. This concept is already proven by atomic clock experiments. The closer an observer is to a gravity source (high mass), the slower time passes for them." (November 7, 2000)

- "By rotating two electric microsigularities [*sic*] at high speed, it is possible to create and modify a local gravity sinusoid that replicates the affects of a Kerr black hole." (November 7, 2000)

- "Personally, I think 'UFOs' might be time travelers with very sophisticated distortion units." (November 7, 2000)

- "It is thought that being close to a gravitational field has a biological effect on all matter including cells. The effect is to slow the movement of electrons in the orbits of their nucleus, which slows the mechanical and biological functions of the observer close to the gravity. Thus the passing of time is a local phenomenon depending on how close you are to a gravitational source." (November 17, 2000)

As we are all too painfully aware, all good things must eventually come to an end, and on March 21, 2001, Titor announced to the world that he would be leaving our time line, and returning to sometime in the year 2036. And then, just as quickly as he appeared, he disappeared. To this day, the mystery of John Titor continues. Who was he? Why did he return and post on public message forums? Wouldn't his postings create some type of paradox?

Titor also faxed information to legendary radio host Art Bell, of *Coast to Coast AM*, in 1998 and discussed everything from how time travel was invented in 2034—"off shoots of certain successful fusion reactor research allowed scientists at CERN to produce the world's first contained singularity engine." He described time as "connected lines and when you go back in time, you travel on your original time line, and when you turn the singularity engine off, a new timeline is created due to the fact that you and your time machine are now there." Titor also predicted Y2K would be a disaster, with people dying and martial law instated and all sorts of horrific events that we all know did not come to pass—at least on *this* time line. Nor did his prediction of civil war in the year 2005 come to fruition.

Bell engaged Titor in questions, which Titor responded to, at one time even telling Bell that his own radio program might be "invaluable to upline researchers" and warning Bell to upload programs concentrating on military technology and new physics theories. "Transcribe these programs and put them someplace safe away from the box. I recommend someplace in the mid west." When Bell asked what the box was, Titor responded with vague references to a war between the U.S. government and that of Russia.

We may never know whether the story of John Titor was true— or simply an elaborate hoax crafted by an enterprising individual for his own personal amusement. In the later part of 2003, a book was released by the John Titor Foundation, Inc. titled *John Titor: A Time Traveler's Tale*, continuing the hype, along with plenty of help from the Internet. Despite many attempts to silence believers by serious researchers tracking down the clues to his identity, including the attempts of Mike Lynch, a private detective who was asked to track Titor down for an Italian documentary after suspicions arose that the lawyer who established the foundation, Lawrence H. Haber, may have been involved (some suggested Haber's brother John, a computer scientist who was found to have a post office box linked to the Foundation, may have been the real Titor), the legend lives on. Most people write it off as a hoax, but Titor has gained his share of followers, who insist there was more to the story than met the eye. Maybe in time we will find out. Whoever he was, he engaged people in an online dialogue for months that crossed the boundaries between fantasy and reality for everyone involved—except Titor himself, who knew the truth about his identity even if the world wondered.

Another possible Titor is a man by the name of Marlin B. Pohlman, an American scientist and author who filed for a U.S. patent application for a time machine in 2004. The patent, which bears application number US 2006/0073976 A1, is titled "Method of Gravity Distortion and Time

Displacement" and is a complicated, diagram-filled "time machine" that can be viewed online at the United States Patent and Trademark Office Website, for those interested in helping Pohlman create a prototype. Is this the real Titor? Who knows? Or is this just a man with a plan involving a lot of math and terminology that would give most people a migraine? Time, again, will tell if Pohlman's device will work.

More Urban Time Travel Legends

Andrew Carlssin—In January 2002, a man by this name was allegedly arrested for SEC violations involving high-risk stocks. The story goes: Carlssin started out with an initial $800 and ended up with $350,000,000 in just two weeks, which got the attention of the FBI and the SEC quick. Once arrested, Carlssin confessed that he was a time traveler from 200 years in the future. He plea-bargained for a lesser sentence by promising to reveal the whereabouts of Osama bin Laden and give up the secret cure for AIDS. All he wanted, it seemed, was to get in his time travel machine and go home. Blame the *Weekly World News* for this one. It originated the purely fictional story, which later got picked up by mainstream media outlets online and in print, and a legend was born as more and more details were added on to the original story.

David Lang—Imagine coming from the year 2550 on a mission, and then making the terrible mistake of falling in love and deciding not to go back to the future. Thus was the fate of one Mr. Lang, who tried to avoid the inevitable trip back to the future by destroying his time machine. Instead, he apparently vanished into thin air on his farm, leaving behind a wife and two kids.

The Chronovisor—Imagine being able to peer into the past and future. That was the claim of one Father Francois Brune, author of books about the paranormal and religious anomalies, who stated that an Italian scientist and priest named Father Pellegrino Maria Ernetti built this device. Although Ernetti did really exist, the actual chronovisor has never been seen.

Interestingly, Brune mentioned the device in his book *The Vatican's New Mystery*. (Sounds like a great publicity stunt to these authors! Perhaps we should have thought of inventing our own time machine device to sell more copies of this book you are holding in your hands, but something tells us our publisher wouldn't have appreciated the deception!)

Modern Dude at 1941 Bridge Opening—Take a look at this photo from the 1941 re-opening of the South Forks Bridge in Gold Bridge, British Columbia and you will see what appears to be a very hip dude amid the normal folk. His sunglasses and clothing are not at all the clothing worn in the 1940s. This photo actually originated in a virtual exhibit called "Their Past Lives Here," courtesy of the Virtual Museum of Canada. Once the Internet went viral with the photo, conspiracies abounded of the hipster dude's

Figure 7-2: This modern dude appeared in a photograph of a 1941 bridge opening. Was he a man out of time?

time traveling identity. Unfortunately, some savvy researchers found out (spoilsports!) that the sunglasses and clothing sported by the out-of-time-and-place man were actually quite readily available in the 1940s and the mystery balloon was proverbially popped. Another one bites the dust.

A Time Traveling "Chair"?

Perhaps not surprisingly, Titor is not the only self-proclaimed time traveler that has appeared and subsequently caused a sensation. We wrote in Chapter 3 about the alleged Montauk Project. But according to some conspiracy theorists there was an offshoot program involving "the Montauk chair," which would supposedly facilitate time travel in the most unusual of ways. Among the "Montauk boys," or those human guinea pigs who had the opportunity to experience this strange device, was one Andy Pero, who was born in Nevada in 1969. His father was a lieutenant commander at the Fallon Naval Air Station in Nevada, where Pero claims he was subject to horrible torture sessions and mind-control programming as part of another project, Superman, intended to create super soldiers.

But it was his involvement in the Montauk Project that allowed him to experience a bizarre time travel "chair" and then enter a wormhole he claimed was large enough for a truck to drive through. In an interview for the noted conspiracy Website educate-yourself.org, Pero talked about how the chair operated. By separating the mind and body, the chair operator's thoughts about a specific time period and vibrational energy would be picked up by a sensitive antenna above the chair, sent to a computer then to a processor, and then amplified several hundred times. According to Pero, this amplification process would create a wormhole and the person would then have access to that time period.

According to Pero, the location at Montauk where this was carried out had special properties itself: "In Camp Hero, Montauk, the location is the cross hairs of the earth's biorhythms and is the point on earth where time travel is most easily accomplished when the earth is the point of origin." He also talked about one major mission he was sent on:

> One such time travel mission was called Project Southern Cross. It was used to win WW2 in favor of the allies. What the US government did using time travel was to go back in time to the 1940's to help us win the war. We would deliver communication devices, weapons and technologies made out of 1940's parts. These would be delivered to the 1940's along with a complete set of drawings on how to make them out of 1940's parts. I took part in several of these deliveries, one time I was sent to Germany and another time to England. I was not allowed to speak to anyone, other than deliver my parcel and quickly return back to our time. And this was all done under deep hypnotic programming, so I didn't have a lot of freedom to explore. I was gone no longer than two hours for the deliveries....

Pero also talks about a mysterious "monster" that was materialized by someone named Duncan Cameron, using the Montauk chair.
Interestingly, Cameron is the brother of Al Bielek, who has his own
connections to the Montauk Project and time travel as an alleged first-
hand participant in the Philadelphia Experiment, the granddaddy of
time related conspiracies (covered in Chapter 3 of this book). Most
researchers have debunked the Philadelphia Experiment as well, but
remember: Where there is smoke, there may be a hint of fire, as we
now have the technology to develop "cloaking" abilities involving both
space and time.

Through the Looking Glass

Pero isn't the only person talking about his involvement in
government/military operated secretive time travel hijinks. Dr. Dan
Burisch, who claims to have once worked for the ultra-top-secret Majestic
program as well as have connections with the mysterious Groom Lake/
Area 51, aliens, and alternate time lines, discussed his work on-line
(*danburisch.info* and *projectcamelot.org*) with Project Looking Glass, alleged Stargates, and even time travel in the form of manmade devices
that can alter and shift time lines. To read the interviews Burisch has
given is to enter the world of science fiction and fantasy, yet he and others, such as Bob Lazar, a UFO enigma in his own right who claimed to
have witnessed the back engineering of crashed alien craft at Area 51,
state with conviction that these things indeed are being experimented
with at Groom Lake.

Other "inside contacts," who go by mysterious names such as
DonDep, corroborate Burisch's claims about stargates or "rings," which
are manmade devices (there are natural stargates as well, one of which
is said to be at Sedona, Arizona) that have appeared in several places,
including in Iraq and have apparently been used by dictators, including
Saddam Hussein, for the purposed of linear time travel. In an interview
on Burisch's own Website, he states:

> The natural ones, like the one at Frenchman, Sedona, etc. are
> various sized, depending on whether they are being opened by
> natural (cosmic) or by unnatural (electromagnetic pressure)
> means. The manmade one's [*sic*], such as Alice's Looking Glass,

was about 20 or so feet in diameter if I am remembering exactly right...I only saw it a couple of times.

These men claim that these rings pose a grave threat to the destiny of the planet because of the consequences of their use, or misuse, by both those with evil purposes and those who just wish to understand the science behind them. Only a few people currently know how to negotiate the machine and its time lines, and according to Burisch "those people are well secured." He also claims that the Timelines, of which there are two, will interface in 2012 (were the Maya onto this?). Scary stuff—but is it true?

Claims such as these, involving alleged Black Ops programs by folks with an inside track to the Illuminati and HAARP and discussions of Timeline 1 and 2 and who is living in which one, are hard to prove or disprove, and that is part of the mystique of such theories. *What if?* What if these people are telling the truth, and because it sounds so outlandish, we automatically pass it off as fiction? But also we must ask: What if these people are lying or creating scenarios for motives and agendas only they understand? That is the allure of conspiracy, and the magnetism of secrets. We know our government and military entities do conduct programs outside the visible boundaries of the average human's perception, so to speak. Because of that, anything can be claimed or alleged to occur, without ever having to have solid proof. The secretive nature of the program cannot, by its own very secretive nature, be "shown" to the public. We can believe, then, whatever we choose to believe, and no doubt find some kind of "circumstantial evidence" to back up our beliefs.

Many time-altering stargate claims abound on the Internet. A quick Web search revealed sites claiming threats of total destruction or the complete evolution of the human species (such extremes!) on December 21, 2012 (at least you have time to read this book!), an enduring battle between military and government entities over stargate control, claims of ancient alien races of beings that control the technology to make these stargates, and even natural stargates in the form of Einstein-Rosen Bridges like the one in the "Burmuda Triangle (sic)," as one site put it. Apparently, today's theoretical physicists who are still grappling with proving wormholes exist beyond a shadow of a doubt (not to mention how to make a wormhole traversable without killing everything that comes near it) are way behind the times. It's already been done! One claim alleges that the entire Iraq war was designed to recover a stargate that was unexpectedly located there. Discernment and critical thinking is mandatory when dealing with the wild world of conspiracy.

Chrononauts Among Us?

Another intriguing claim of time travel via the government comes from a man named Andrew D. Basiago, who claims as a child to have been a participant in Project Pegasus, a classified, defense-related program under DARPA to perform what he calls "remote sensing in time" experiments, which would provide the intelligence community with information about the past and the future. Apparently, 140 schoolchildren were involved in the project, which Basiago speaks about in a series of YouTube videos and ample online interviews, and that these children became the nation's first generation of "chrononauts," or time travelers.

Basiago claims to have teleported as a child, and today serves, interestingly, as the new Project Pegasus team leader, lobbying the U.S. government to declassify secret time travel and teleportation documents. On his Website (*www.projectpegasus.net*) he refers to himself as a "21st-century visionary" and an emerging figure in a truth movement designed to reveal what the government is covering up about life on Mars and "quantum access" to the past and future, among other things. He also founded MARS, the Mars Anomaly Research Soceity. He has in impressive background to be sure IMENSA status, five degrees including a BA in history from UCLA and a master's of philosophy from Cambridge University), and he has a long and illustrious career as a journalist, not to mention other achievements. But his claims of being one of the first time travel guinea pigs? Those are harder to prove and, until the actual proof does make itself manifest to the public, will continue to keep Basiago's story, even if completely true, relegated to the fringes of research (where, let's remember, pieces of the truth often reside!).

Those who have talked to and heard Basiago speak are convinced that his experiences are real and that one day we the people will know the truth about DARPA's time travel experiments, just as Burisch no doubt believes all will be revealed about the stargates and Timelines few know exist. Recently, in the *Seattle Examiner*, a physicist named Dr. David Lewis Anderson was said to step up as a second whistleblower, confirming Basiago's claims about Project Pegasus. Anderson, director of the Anderson Institute, claims the U.S. Air Force was involved in time control research, which he himself went on to continue at his "Time Travel Research Institute" in Long Island and other such organizations. He was, as well, a young participant when the Air Force began conducting its time related research at the Air Force Flight Test Center at Edwards Air Force Base in California's Mojave Desert. He claims the research was called "time-warp field theory" and modeled "how to use the natural forces of inertial frame dragging to create contained and controllable fields of closed time-like curves." He also claimed research was investigating "quantum tunneling, time-warp fields, and wormholes."

Interestingly, the focus on quantum physics and time travel, as discussed in earlier chapters, is a natural for any investigation into time travel itself, so these experiments shouldn't come as such a shock. No doubt, if our scientists are putting forward amazing theories on how we might go back in time or forward into the future, there will be agencies hard at work developing the means and technologies to do so. So perhaps these "conspiracy theories" are not as hard to embrace as they may seem upon first being exposed to them, and maybe these men (and women) who were participants truly did get an inside glimpse into the race to build the first time portal, stargate, or device.

The credibility problem will not go away until there is full disclosure, and even then many will choose to label the disclosed information as complete BS, which creates a catch-22 we cannot often escape (damned if you do, damned if you don't), but the problem with making claims that are boundary-pushing is this: The more outlandish they are, the more credibility you lose. This is why the tin foil hat jokes were flying in early January 2012 when the report hit the Internet that our own President Barak Obama was himself a Project Pegasus "chrononaut," or time traveler. Basiago and another fellow chrononaut, William Stillings, revealed to the press and public that Obama was one of their own, a participant in the DARPA project under the name of "Barry Soetero." Obama, they claim, was 19 at the time and was one of 10 young people to teleport to and from Mars as part of the program that occurred in a top-secret "jump room." Obama, they claim, has not only lied about his identity, but he has concealed his involvement in secret CIA backed "Mars" training classes at a California college in 1980. Basiago and Stillings identified the current head of DARPA, Regina Dugan, as another fellow chrononaut and Mars jumper. The purpose of these Mars trips was to "acclimate Martian humanoids and animals to their presence," Basiago stated (in Alfred Lambremont Webre's article on Examiner.com, "Second Whistleblower Emerges to Confirm reality of Time Travel").

The best response came from Tommy Vietor, National Security Council spokesperson, in the January 2012 "Danger Room" column on Wired.com, who stated that Obama never went to Mars: "Only if you count watching Marvin the Martian." And of course, conspiracy buffs all over the world immediately suggested that Obama and the entire government were lying through their teeth, and that every claim was true. Further still, Basiago claimed in a 2011 interview with George Noory on Coast to Coast radio that Obama, George Bush Senior and Junior, Bill Clinton, and Jimmy Carter were "pre-identified by Project Pegasus as future U.S. Presidents." This information was discovered via "quantum access," he claimed, but also that the prediction of Carter was gleaned from a book that wouldn't be written until many years later in 2005, notably the Universe Books edition of *Exopolitics: Politics, Government and the Law in the Universe* by Alfred Lambremont Webre. Supposedly, other such books were retrieved from the future by DARPA and supposedly returned to 1971 (or possibly even earlier) to be studied. Mr. Webre, apparently, is now under "time travel surveillance" by the U.S. government.

How can you argue with predictions that are revealed after the fact? Perhaps if this information had slipped out when it was first learned, we might still have a surplus and not a massive deficit to deal with!

Maybe we got some of this amazing time travel technology from aliens, if you believe a conspiracy involving one of the greatest UFO cases ever—Roswell.

The UFO Connection

In a nutshell (because the Roswell story could take up several books itself), in the summer of 1947, something crashed on a ranch in the desert near Roswell, New Mexico. (See Nick Redfern's sidebar on page 151.) That "something" is, to this day, still being argued by those that were witnesses, those who knew the witnesses, and those who have studied this granddaddy of UFO crash cases for decades.

The status quo explanations run the gamut from "crashed alien UFO with occupants that were later hustled off to a top secret base and the craft taken to Area 51 for back engineering" to "a simple weather balloon that was misidentified." Of course, conspiracy abounds, but even time travel managed to worm its way into this amazing case. One such story involves a renowned U.S. Navy Commander during the 1950s named George W. Hoover, who revealed to a select group of people, including his son, George Hoover, Jr., in the 1960s and later to researcher William J. Birnes that the Roswell aliens were not from Mars or even aliens, really, but were in fact "not so much interplanetary as much as they were literally also time travelers," according to an article by Andrew Bragalia in *The Ufo Iconoclasts*. He called them "extra-temporal." Hoover also admitted his own work as a Naval Intelligence Officer with top-secret clearances and his access to the Roswell crash "secrets," as well as the purpose of these visitors he claimed were clearly from the future. He even alluded to Birnes, in an interview, and to his own son that these entities may be "us" form the future—our future selves. Hoover also stated that the government feared the intentions and abilities of these time traveling visitors and that the visitors were able to use the power of consciousness to morph reality, something we humans are also capable of, but do not realize.

Hoover's amazing story wasn't the only mention of time travel with regard to Roswell, as Nick Redfern suggests here on page 152. Many UFO contactees, who have established regular contact with alien entities, also report time travel aspects and information imparted to them. Billy Meier, one of the most well-recognized "contactees" was a farmer living in Switzerland who began experiencing visits in 1942, at the age of 5, that lasted decades, from an elderly ET human man who called himself Sfath, and later with ET/human women named Asket and Semjase. Meier is known for his vast collection of photographs, controversial to this day, and his stories of beamships and communications with these beings from the Pleiades, earning him a large following of "believers." One of Meier's claims involved traveling back in time and meeting someone named "Jmmanuel," who he claimed was the real Jesus Christ.

Anne Strieber, wife of Whitley Strieber, whose own amazing contact experiences with entities has been documented in a series of books and movies, such as *Communion* and *Transformation*, writes on the Unknown Country's (*www.unknowncountry.com*) "Insight" news column about the possibility that the "alien" visitors that many contactees may be communicating with may be time travelers here not from another planet, but from another time. They may even be coming into our time frame to keep an eye on us—even warn us—of pending nuclear proliferation and global climate change. The late John Mack, who worked extensively with contactees, also pointed to messages from entities about saving the planet and ecological doom. Maybe they are indeed the future "us," as George Hoover told Birnes and his son, trying to keep the present "us" from messing up our own future!

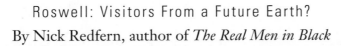

Roswell: Visitors From a Future Earth?
By Nick Redfern, author of *The Real Men in Black*

*Figure 7-3: The actual site from where the Roswell "UFO"
wreckage was recovered.*

It was only a little more than a week after Kenneth Arnold's famous June 24, 1947, encounter of the Flying Saucer kind at Mt. Rainier, on Washington State's Cascade Mountains, that a highly unusual aircraft plunged to earth in the deserts of Lincoln County, New Mexico, not far from the now-infamous town of Roswell. The deeply controversial event has been the subject of dozens of books, official studies undertaken by the General Accounting Office and the U.S. Air Force, a plethora of television documentaries, a movie, and considerable media scrutiny and public interest.

The admittedly weird affair has left in its wake a near-mountain of theories to explain the event, including a weather balloon, a "Mogul Balloon" secretly utilized to monitor for Soviet atomic-bomb tests, an extraterrestrial spacecraft, some dark and dubious high-altitude-exposure experiment using Japanese prisoners of war, an atomic-based mishap, the crash of a Nazi rocket with shaved monkeys on-board, and an accident involving an early "Flying Wing"–style aircraft, built by transplanted German scientists who had relocated to the United States following the end of World War II.

It is no secret that I'm distinctly skeptical of the idea that aliens met their deaths in the desert on that long-gone day in July 1947, and I consider that should we one day uncover the true story of what really occurred outside of Roswell, it will likely be one of secret military experimentation born out of the early years of Cold War shenanigans. But, what if the Roswell affair is explainable in a very different, and wildly alternative, fashion? What if the craft and its strange crew were not the denizens of another galaxy, or of the U.S. military of the immediate post–World War II era? What if, incredibly, their point of origin was a far-flung future of a distinctly human nature? Though such a scenario may sound extreme to many—even to those who are of the opinion that something truly anomalous occurred at Roswell—such theories have been expressed, and endorsed, too.

One of those who revealed his thoughts on this particular scenario was Lieutenant Colonel Philip J. Corso, co-author with William J. Birnes of the much-debated 1997 book The Day After Roswell. The sensational story told of Corso's personal knowledge of the Roswell affair while serving with the military, and of the way in which he allegedly helped to advance the United States—both scientifically and militarily—by secretly feeding certain fantastic technologies found in the craft recovered at Roswell to U.S.–based private industries and defense contractors.

Despite the fact that many have championed Corso as a solid proponent of the idea that extraterrestrials plunged to earth in New Mexico in 1947, in reality, Corso was willing to consider something very different. The unusual bodies found within the wreckage of the craft, Corso explained, were genetically created beings designed to withstand the rigors of space flight, but they were not the actual creators of the UFO itself. Right up to the time of his death in 1998, Corso speculated on the distinct possibility that the U.S. government might still have no real idea of who constructed the craft or who genetically engineered the bodies found aboard.

Notably, Corso gave much consideration to the idea that the Roswell UFO was a form of time machine, possibly even one designed and built by the denizens of an Earth of the distant future, rather than by the people of a far away solar-system. Maybe, by studying the Roswell materials, officialdom has learned something deeply troubling about our future, something it dares not share with us, the populace at large. Is this, perhaps, the reason why the Roswell affair is one still shrouded in overwhelming secrecy, more than 60 years after it occurred?

To paraphrase The X-Files, when it comes to UFOs and Roswell, "the truth" may not be "out there" after all. Instead, it might be countless millennia ahead of us....

Of Birds and Baguettes

Sometimes, a conspiracy can have a very real origin and where there is smoke, there really is some fire—as in the case of the now-famous "Baguette Conspiracy"! On November 3, 2009, the super cooled magnets at Sector 81 of CERN's Large Hadron Collider began to mysteriously overheat, threatening to cause some serious damage to the sensitive equipment. Scientists scrambled to find out what was causing the problem and eventually found the culprit: a piece of baguette! Somehow a chunk of bread got dropped onto an electrical substation located above the accelerator, which caused the problem. They got rid of the bread and returned the system to the cryogenic state needed to operate.

Two physicists who were not satisfied with that mundane explanation came up with a theory of their own: that a time traveling bird had somehow been sent from the future to sabotage the experiment! These were not two hacks, but brilliant physicists, namely Bech Nielsen of the Niels Bohr Institute in Copenhagen, and his Japanese colleague, Masao Ninomiya of the Yukawa Institute for Theoretical Physics.

The two physicists had published a series of papers prior to the baguette incident, suggesting that the experimentation going on at the LHC was purposely sabotaged by visitors from the future because it was deemed unacceptable to the universe to tamper with such things they found "abhorrent to nature," such as the Higgs Boson, which the men felt was not yet to be found because each attempt was somehow rejected by the universe because of the implications it would have. The future, they stated, would simply intercede to prevent the success of the experiments in the present day (sounds like a reverse-paradox to us!).

Although their theory did have plenty of math to back it up, it really remains, to most of their colleagues, a non-issue, because what is going on at the LHC amounts to nothing more than trying to re-create things that already occur in nature. And one can always ask: If God or the Universe or nature didn't want any of this activity to occur at the LHC in the first place, God or the Universe or nature would not have allowed for it to have been built. Nor would God have allowed for the physicists involved to have ever graduated university and gotten jobs at the Collider!

■■■■■■

We asked nuclear physicist Stanton Friedman, one of the most widely respected UFO researchers and lecturers around, about his views on UFOs and time travel. He is the author of the best-selling books Flying Saucers and Science, Crash at Corona *(with Don Berliner), and* Science Was Wrong *(with Kathleen Marden).*

Time Travel

By Stan Friedman

(*www.stantonfriedman.com*; fsphys@rogers.com)

Considering that since 1967 I have been lecturing about flying saucers, as a nuclear physicist, in all 50 states, 10 Canadian provinces, and 18 other countries, and have appeared on hundreds of radio and TV programs, it is not surprising that I have been asked my views about time travel as an explanation for how anybody might get here. This is usually the result of a belief system assuming that civilizations are very few and far between in the galaxy and that there are no conventional techniques that could be used. Just pop in here and come out there. It happens in science fiction fairly often.

As an example, Dr. Frank Drake of the SETI (Search for Extraterrestrial Intelligence) Institute, using his famed Drake equation to determine (guess would be a better word) the number of civilizations in the galaxy, concluded there might be as many as 8,000 planets from which radio signals might be sent. He has repeatedly stated that interstellar travel was impossible and, of course, neglected the possibility of colonization and migration, which are the techniques used to distribute intelligent life around planet earth. Drake's colleague, Dr. Jill Tartar, a few years ago was rhapsodizing that there might be another civilization within 1,000 light years of the sun. In 1522 Magellan's sailing ship took about three years to circumnavigate Earth. Jules Verne cut that down to 80 days in the 19th century using a balloon. Today the space station and many other orbiting satellites go around the earth about every 90 minutes. Of course that they do so is to be expected based on the basic rule for technology growth, namely that technological progress comes from doing things differently in an unpredictable way.

The astronomical and SETI communities seem to think that chemical propulsion is the end all for space travel. Their thinking is stuck in the past. A rough analogy would be to assume that the biggest bombs we could build would release the energy of 10 tons of dynamite as did the10-ton blockbusters used to devastate Europe in 1942 and 1943. Of course, in 1945 we exploded our first atomic bomb, releasing the energy of 19,000 tons of TNT. Only seven years later our first hydrogen bomb (nuclear fusion device) was exploded, releasing the energy of 10,000,000 tons of TNT. A Soviet H bomb released the energy of 57,000,000 tons of TNT a few years later. This may seem far removed from star travel.

But consider that the United States Navy launched its first nuclear-fission-powered submarine, the *Nautilus*, in 1956, capable of remaining underwater while going around the planet. Earlier chemically powered subs could remain underwater only for a day or so because of the need for air for their propulsion systems. Hundreds of nuclear-powered submarines have been built since. Perhaps most impressive are nuclear-powered aircraft carriers, which can operate for 18 years without refueling. A number of studies have examined the use of nuclear fusion for deep space propulsion and found that, using the proper hydrogen and helium isotopes, charged particles could be ejected from fusion rockets in deep space. They would have 10 million times as much energy per particle as we can provide in a chemical rocket. Hydrogen and helium are the two lightest elements and the most abundant in the universe. Uranium, the basis for fission bombs and nuclear-powered ships, is relatively rare and is almost twice as heavy as lead per unit volume. Los Alamos National Laboratory, Aerojet General, and Westinghouse Astronuclear Laboratory operated even so successful nuclear fission rockets in the late 1960s. The Phoebus 2B, built by Los Alamos, was less than 8 feet in diameter but produced 4,400 megawatts of power, or twice the output of Hoover Dam, which is obviously very much larger. The primary use of such systems would be for upper stages in rockets used to settle the moon or Mars. The cost of such programs would be huge remembering that *Apollo* cost $20 billion.

It must be noted that as far as we know there has only been advanced technology used by Earthlings since about 1900, although the Earth is at least 4+ billion years old. Unfortunately there are still people who accept the strange notion that the Earth was only created in 4004 BC.

One of the important aspects of this discussion is the notion that civilizations in our galaxy would be very far apart from each other. That has changed drastically because of the huge number of measurements made by the Kepler space satellite. The recent estimate by those scientists is that there are probably one hundred billion planets in the galaxy. We also know that there are more than 1,000 stars within a mere 55 light years of the sun. Some are unquestionably much older than the sun and some most likely have neighboring stars much closer than the sun is to Alpha Centauri, 4.3 light years away.

Exciting examples of a very interesting system are the two stars Zeta 1 and Zeta 2 Reticuli. These two sun-like stars are only 39.3 light years from here (the galaxy is about 80,000 light years across) and are a cool billion years older than the sun. In addition, instead of being relatively isolated like the sun, they are less than an eighth of a light year apart from each other, or more than 30 times closer to each other than the sun is from Alpha Centauri. Each can be observed easily all day long from the other. The real difficulty for the SETI community is that the discovery of this special pair is the result of a major effort made by Ms. Marjorie Fish, stimulated by the Star Map drawn by Betty Hill based on what she was shown by aliens on-board a flying saucer on September 19, 1961, in New Hampshire. It is a basic rule of the SETI and astronomical communities that one must ignore all data connected to flying saucers. They consistently ignore the large-scale scientific studies of UFOs, as well as the advanced travel technology research and development being done in industry and the national labs and not in academia. Dr. Seth Shostak, a prominent member of the SETI community, has even suggested that, because they have used their radio telescopes to listen for a signal from Zeta 1 and Zeta 2 and not found any, there must be no intelligent life there! Of course, they haven't found any intelligently produced radio signals from any other planet around any other star (except the Earth), and they have not provided any evidence that ETs are stuck at our primitive level of radio technology and would have any reason to send a signal to Earth, which has a very primitive society whose major activity is obviously tribal warfare and which sent its first "long distance" radio signals only 112 years ago. They would have launched their Kepler a long time ago and would know there is at least one planet in the Goldilocks zone around the sun. The star map work is discussed in detail in a chapter in *Captured! The Betty and Barney Hill UFO Experience* by Kathleen Marden, Betty's niece, and me. Advanced technology for star travel fills a chapter in *Flying Saucers and Science*, my 2008 book. There is a 10-page bibliography, including 10 UFO related PhD theses, in my *TOP SECRET/MAJIC*.

The reason for this extended discussion is that many people believe that ET life is thinly distributed and is too far away for any known travel technology system and, therefore, time travel with wormholes and exotic physics must be the solution.

We of course do know another means for time travel, moving at close to the speed of light. Einstein showed, and it has been demonstrated for particles, that as one gets close to the speed of light, time slows down for the things moving that fast—crazy as that might sound. It must also be noted that at just 1G acceleration, it only takes a year to get close to the speed of light. At 99.9 percent of the speed of light, it only takes six months pilot time to go 37 light years. So go off, and come back and marry your grandchild's best friend! This is real, not science fiction, and has been demonstrated with both large accelerators and cosmic rays. None of this is to say that others elsewhere have not learned how to do time travel—only that it isn't necessary for there to be contact between neighboring solar systems. Just notice how fast our technology has evolved since 1900, and then imagine where it will be, presuming we survive, a thousand or million years from now.

Maybe my problem is that I am not a theoretical physicist and, having worked on a number of large highly classified programs, am much more aware than most of how much information about flying saucers, alien visitors, and advanced technology has been hidden, I will listen when the theorists provide evidence of string theory, space time warping, and time travel.

We are, we regret, out of time for this chapter, but wanted to leave you with this: Whispers of the truth are everywhere, and it is up to you, the discerning reader, to figure out what that truth may be. We cannot do it for you. All we can do is provide you with the means to examine the smoke that may—or may not—lead to fire.

8

LET'S DO THE TIME WARP AGAIN

We all have our time machines. Some take us back, they're called memories. Some take us forward, they're called dreams.
—Jeremy Irons

Your brain thinking back into memories is technically time travel in your brain.
—Max Jones, age 11

Time has no divisions to mark its passage, there is never a thunderstorm or blaze of trumpets to announce the beginning of a new month or year. Even when a new century begins it is only we mortals who ring bells and fire off pistols.
—Thomas Mann

Physical time travel in a machine eludes us, but does that mean that other types of time travel may not exist? It may be decades—even centuries—before we find ourselves zipping off to the year 1209, or 2901 if you prefer, in a designer Tardis, but maybe we time travel already in ways we don't even realize. Dreams. Memories. Altered states of consciousness. And maybe there are places on Earth where time doesn't always behave itself—places where we experience what are called time shifts, time slips, and time anomalies.

This chapter gets a little "paranormal," but we cannot leave this subject of time without looking at all the ways it presents itself as a part of the human experience. Ask someone who has had a precognitive dream about an event that came true, as dreamed, three weeks later, and he will tell you: He traveled forward in time while asleep. Ask someone who experiences a powerful déjà vu with past life recall of a location she knows she's never been to, and she will insist she was at that same place—thousands of years ago.

We will be featuring many experiential stories in this chapter of other ways people may experience time. Unfortunately, though, it isn't always simple, linear, and neatly packaged, as our normal waking brains would prefer it to be.

Precognition

Precognition is the ability to see into the future, often for the sake of retrieving important warnings or information that may help make a present choice or decision to avoid such an outcome. Psychics and others who possess psi abilities have long posited that they have the ability to access information outside of the limitations of linear time, and, believe it or not, even science points to the possibility of such a field of "timeless" information—the Zero Point Field.

Whether you call it a sixth sense or temporal remote viewing, the idea that we can see into the future has been the subject of many prophecies and predictions that have gripped humanity with fear. Remember Y2K? But the ability also has helped those who've experienced it predict potential disasters they were later able to avoid. Many people reported that on the morning of the September 11th terrorist attacks, they had dreamed the previous night, or perhaps even a week or two prior, of something awful happening—something that prevented those very people from getting on a plane that morning or going into the city near what is now Ground Zero. Were these people accessing the future in a form of mental time travel? Can we even do this in dreams? Are prophetic visions a form of mental time travel?

Science has long attempted to find ways to "prove the paranormal" and usually meets with great skepticism. This happened in January 2011 when an article was published in the American Psychological Association's *Journal of Personality and Social Psychology* by a very well-respected Cornell University emeritus professor of psychology, Daryl Bem. Bem's paper was titled "Feeling the Future: Experimental Evidence for Anomalous Retroactive Influences on Cognition and Affect." In the long and detailed paper, Bem provides a compelling argument for psychic abilities that appear to be able to rise above the confines of linear time and allow the future to reach backward to influence the past. The paper claimed that, in a sense, people like you and me can be altered by things that have not even happened yet, and that time is "leaking" in the form of the future slipping into the past and present.

Bem performed a variety of experiments over the period of eight years to try to prove his theory, which you'll recall has its equal in the photon delayed choice experiments of John Wheeler and other quantum physicists. Bem's experiments often involved randomly chosen computer images that a group of students would be asked to view. Half would be viewing a blank image, the other half a hot, erotic image. One might expect out of 100 sessions there would be a potential 50-50 hit rate, but Bem's sessions often exceeded that when it came to the erotic pictures—almost as if the brain "knew" there was going to be a hot image being shown before it actually was. The students reacted physiologically to the hot images approximately two to three seconds *before* the image actually was chosen.

In another study, Bem took 100 college students and showed them a list of 48 common nouns. The nouns were flashed on a computer screen at a rate of three seconds each, with instructions to look at the word and then visualize before moving on to the next word. Then the students were given a memory quiz to recall as many of the words as they could. They typed these words, and then the computer chose 24 of these words at random. The students then had to try to scan and type the 24 words the computer selected, from memory.

Afterward, Bem was surprised to find that on the original recall test, before the 24 chosen words, the students proved to be better at recalling the actual words they had scanned and retyped *after* the test. The second group of 24 random words was not recalled as often. Bem's conclusion was that practicing a set of words after the recall test could reach back in time to facilitate the recall of those words. Scanning and retyping them *later* improved recall *earlier*.

Critics have jumped all over Bem's research, and attempts to re-create the same outcomes have failed, but other scientists have rallied behind the paper's findings if only to demand more research and serious study of retroactive influence by, as the paper's abstract states, "time reversing well-established psychological effects so that the individual's responses are obtained before the putatively causal stimulus event occurs." Causality, again, is called into play, but we ask you to keep an open mind and remember the photons that, with a change of choice of the observer, changed the outcome of their destiny. The quantum may mirror the brain's own ability to play with time, reverse it, predict it, and change and rearrange it. Maybe it's all just a matter of mental time traveling.

Don't believe it?

"Mental" Time Travel

In December 2010, the online physics community Phys.org published a story about a research study that examined "chronesthesia," or the brain's ability to travel back and forth in time mentally. This activity, scientists say, allows us to remember the past and envision the future, and to "mentally travel in subjective time." Those were their words, not ours! The researchers included Lars Nyberg from Umea University in Sweden, Endel Tulving from the University of Toronto, Ontario, and colleagues who determined that mental time travel consisted of two independent sets of processes: those that determine the contents of any act of travel, such as what happens, who the "actors" are, and where the action occurs, as if you are watching a movie on a screen; and those that determine the subjective moment in time during which the action occurs, such as past, present, or future.

Tulving stated that, in cognitive neuroscience, "we know quite a bit (relatively speaking) about perceived, remembered, known and imagined space. We know essentially nothing about perceived, remembered, known and imagined time." The question they hoped to answer is: When you recall something that you did, say, the previous night, you are not only aware that it happened but that you were there observing it at the time it happened—but how do you know that it happened at a time other than "now"? They were able to pinpoint specific brain activity was different when it involved thoughts about the past and the future as compared with the present, but that brain activity was very similar for thoughts about all of the non-present times (imagined past, real past, imagined future).

The results showed that mental time is a product of the human brain, and is different from the external time we call "subjective time," measured in clocks and calendars. Chronethesia, then, is a form of consciousness that allows us to think about subjective time and travel mentally in it!

Think about the average time we spend pondering the past or wondering about the future, and it is easy to see how perception of time can shape time, at least in our personal and subjective experience. Think also about how accurate our own memories of our pasts are, and we can see that perception may even construct the way we experience time. We, in fact, use these memories, as inaccurate as they may be, to envision future scenarios of our lives, without realizing that memory comes back to us in bits and pieces, and not usually as the whole picture. We create a narrative based upon those scatterings of memories and impressions of the past, to interpret and respond to our present, and create an anticipated view of our future. And all of it may be based upon things that never really happened the way they did in objective time. Our worldview may very well be structured more upon our experience of subjective time.

Time Dilation in the Brain

Most people will agree that time seems to pass more slowly during certain events and speed by during others. It also seems to go by slower when we are children and have less responsibilities and distractions than as adults, with lives full of activities, worries, and concerns. Is this because the brain is doing something differently, or because our perception of time is just shifted? Many people who work in law enforcement and those in the front lines of battle attest to a shift in time perception involved with very dangerous situations. In one research experiment to test this theory, neuroscientist David Eagleman of Baylor College of Medicine asked a group of volunteers to expose themselves to a dangerous situation. Volunteers had to dive backward with no ropes or chords attached into a net to break the 150-foot drop. They would reach speeds up to 70 mph during the fall. Afterward, volunteers estimated their falls lasted a third longer than the actual falls they witnessed fellow volunteers taking. Eagleman and colleagues also strapped devices that he called "perceptual chronometers" to the wrists of volunteers. These devices, looking very much like modern wristwatches, displayed a series of numbers on the face. However, these could be adjusted in speed so that the numbers appearing were too fast to see.

Eagleman, who described the experiment in the December 11, 2007, edition of LiveScience.com (and later published in the December 11, 2007 journal *PLoS One*) wanted to see if people in a dangerous situation would develop the sort of "slo mo" Matrix perception of time, like Neo did in the famous science fiction movie *The Matrix*, hanging in mid-air as he dodged the bullet. If the brain did indeed speed up in dangerous situations, the scientists thought that perhaps the numbers on the chronometers would appear in slow motion and could be readable to the volunteers. However, the opposite actually occurred: The volunteers could not read the numbers at faster-than-normal speeds. The result, the researchers posited, was a trick of "time warping" played by the volunteers' memory, and that when someone is scared, the amygdala becomes more active and literally sets down an extra layer of memories on top of those normally taken care of by other parts of the brain.

Eagleman said, "In this way, frightening events are associated with richer and denser memories. And the more memories you have of an event, the longer you believe it took." Eagleman also stated that this illusion was similar to the phenomenon of time speeding up, as you grow older. "When you are a child, you lay down rich memories for all your experiences; when you are older, you've seen it all before and lay down fewer memories"—thus, the endless summers of childhood that we adults believe went by in a flash! This is also similar to the sense of slow motion that might happen during an accident, attack, or dangerous situation where the bystander or witnesses report a different perception of how much time passed. To those directly involved in the incident, time seems to slow, and perhaps it even does so to enable the person to think and act faster as a result. It is almost as if the brain is giving the victim of the accident "extra" time to react, when in fact objective time is moving along the same as always. Think of it as the brain's own version of time dilation.

People who meditate regularly report a similar shift in their perception of time, suggesting that simply removing oneself from the distractions of the objective world can create a shift in consciousness that allows time to be experienced differently. In the case of meditators, the present moment becomes the center of attention, so to speak, and thus a deeper sense of experiencing the memories being made in the present result. When we are too busy with life, time tends to speed by and not allow us the chance to breathe. When we are bored out of our skulls, time obliges by dragging on endlessly, and yet, in the act of meditation, the sense of expansion of the present moment acts to not only sharpen the focus on the brain, but also improve memory afterward. Meditators often claim they can retrieve memories more quickly and clearly than those who never slow down long enough to even recognize the moment at hand.

Perhaps even death is a shift in the perception of time.

The Mystery of Time and Death
By Anthony Peake

Time is probably the second greatest mystery of the observed universe. The greatest mystery of all, however, is what happens to consciousness when we die. Up until very recently these twin mysteries have been treated as totally separate challenges. Philosophers, scientists, and mystics have spent centuries debating and discussing both, but never, ever, has it been suggested that one may explain the other. This all changed in 2006 when my first book, *Is There Life After Death?: The Extraordinary Science of What Happens When We Die*. In this book I presented a concept that I call "Cheating the Ferryman," a concept that suggests that immortality is provable by using the very science that for many rationalists have used to preclude the survival of consciousness after the body is seen to die and return back to its inanimate and non-sentient constituents. And the proof may lie in a radical re-thinking of the nature of time itself.

For most people time is the most natural thing in the universe. It is with us continually. Indeed the very concept of a timeless Cosmos is literally unthinkable. Everything happens *in* time and everything needs time to evolve and change. Without time nothing would *happen*. Everything would just grind to a halt. Indeed one of the most profound questions that can be asked is: What happened before the Big Bang? If time was created with everything else at that moment, then how did the spontaneous creation of everything that is, get started? If time was created with everything else, there was not time available for the whole shebang to be started. An explosion needs a cause, and a cause needs time to have an effect. No time, no process of cause and effect. Simple, but brain-numbing in its implications.

But what if time is not what it seems? What if the true nature of reality is a timeless place that exists in a permanent "now"? What if time itself is a creation of the observing mind—an illusion that helps us to release ourselves from the "now"?

This is not such a crazy idea, as it first appears. Most people are aware that time seems to slow down and speed up depending upon mood and circumstance. How many times have we heard people report that during accidents or during a period of extreme stress that time slows down, or that a person experiences a dream that lasts for hours but in "normal" time takes no longer than a few seconds to be experienced? This is because time perception depends not upon the external universe but on brain chemistry and emotion. Indeed, of great significance is the regular reports given by survivors of near-death experiences (NDEs). A large percentage claim that time slowed down for them—slowed down to such an extent that it became almost meaningless as a concept. Why is this? Furthermore others have reported that they experience their whole life in a split second. Clearly something very strange happens to time at the moment of death.

Could it be that, at death, time ceases to be—that we experience for the first time in our life that time is a brain-constructed illusion? Time doesn't exist in dreams, so why should it exist as we approach death? And this is where it gets interesting. Death exists in the illusion that is time. Just like the Big Bang arose spontaneously from a timeless void so at the end of our lives we return a timeless void, a void that is not empty because consciousness does not need time to exist within. Thought does not exist in three-dimensional space and it exists outside of time. Our "inner universe" of pure consciousness does not need a temporal flow to change and evolve. It simply "is."

Death is something that happens in time, at a specific point in the future. But by falling out of time before we reach that location in time and space we never arrive at that destination. We never pay the Ferryman his due to take us across the River Styx to the land of the dead. Charon has been cheated. Time, or more accurately the lack of time, helps us "Cheat the Ferryman"!

Survival Mechanism

Consciousness and perception, we've already seen, have their own ability to shape and distort time for the purpose of adapting to an event or situation, as often experienced in warfare, where soldiers need a sense of time slowing down to have an extra edge of self-protection. Perhaps the ability to shift our way of perceiving time is part of a built-in survival mechanism, and actual time travel is not something we really need to survive. As much as we may want to experience being able to jump into a ship and go back 20 years or forward 50, unless our survival depends on it, it won't happen. Memories and re-experiencing memories, on the other hand, often help us deal with present and future situations by recalling dangers and reviewing mistakes in order to do it better next time—and survive.

Mental time travel also implies, however, having both feet planted firmly in the subjective realm. Yet there are things that people report, all over the world, that suggest that time is being experienced just a little bit differently in both the objective and subjective. Time slips have been reported for centuries, often in association with UFO reports, and even reports of ghosts and apparitions, suggesting that there may be places on Earth where the veil between realities is thin enough to allow us to enter into other realities, and experience that which enters into ours. Often, the experience of a strange fog or mist is present, with no apparent weather-related cause, and a person enters the fog only to somehow come out in another place or time frame. Other reports suggest that people may be experiencing more than one time line at a time, with ample crossover to confuse them. (Could déjà vu be something like this?) Maybe those parallel universes we wrote about earlier really can be experienced now and then, and even at the same time!

Time Slips

In her book *Time Storms: Amazing Evidence for Time Warps, Space Rifts and Time Travel*, British author and researcher Jenny Randles documents dozens of such cases of witnesses entering a strange cloud or fog and being transported through time and space. Often, these people lose hours—even days—of "real time" during their experiences, which leaves them disoriented and confused. These periods of missing time often involve the sighting of a UFO and possibly abduction activity, but not always. Sometimes, people are just bumped up the road a bit, showing up in a town 500 miles away from where they started, or where they intended to be. Sometimes, people see things that are obviously "out of place and out of time," such as ghostly apparitions from the past. Many reported sensations such as ear popping and hair standing on end, and a sense of heavy pressure, and no doubt some electromagnetic and atmospheric anomalies were present to account for the mysterious fogs. These may be similar to the fogs often reported in the notorious Bermuda Triangle, such as the strange green mist that a pilot named Bruce Gernon and his father entered in his own amazing experience in the region on December 4, 1970, when his light aircraft entered a strange lens-shaped cloud that then led the craft into a tunnel-like expanse before emerging into a green mist that clung to the plane and caused the craft's instrumentation to go haywire.

It may even be possible that there exists all over the planet potential Earth-based wormholes or vortices, as some paranormal researchers call them. These are places of high strangeness where anomalous activity occurs more often and people experience distortions of time and space. But these events can occur anywhere, and may or may not be accompanied by UFO sightings and other anomalous activity. Here is one account we received from someone named Rabbi Aaron, who felt he had experienced such an anomaly.

In 1996 I remember that it was Mariel Hemmingway that was dead and not her sister Margaux. I was stationed in England at this time and I remember it being Mariel's name they said on the radio and that she had been possibly murdered. I was very confused when a few years later I saw Mariel on TV and when I investigated further learned it was her sister Margaux that was dead by suicide. Looking back I remember both time lines very clearly but for a few years I only knew the one timeline, if that makes any sense—kind of like in the movie *Frequency* and the son remembers his dad being killed on the job...but when he's able to help his father avoid being killed he remembers his dad as being both dead and alive with the additional memories of a his dad living—that's how my experience is like. The experience of repeated time happened in 2005 and in 2007. I remember going to my therapist and she was documenting this stuff.... I felt like I was going crazy because I knew what was going to happen and no matter what I did I couldn't really alter the events.... I described for her during both years what the events of that year were going to be and they passed as described...but this wasn't just a vision of the future or a psychic event; I felt as if I was physically present during this repeated time line and I was aware of time passing. I still sometimes forget that I am actually two years younger than I really am because I remember living the additional two years. I tell everyone that I'm 39 but then when that's called into question (because I look younger than my age) and I think about it or have to show my driver's license I realize that I'm only 37. It's weird to say the least...and I don't know why this has happened to me but I have a weird feeling it's related to another experience from the evening of September 11, 2001. On that evening I saw a UFO.... I'm not saying it was extraterrestrial; I'm just saying I haven't seen anything like it before. I was feeling compelled to go to it even though I didn't know it was there and I felt like I was there communing telepathically with the occupants for a couple hours but it really had only been 42 minutes from the time I left the house and returned and 21 minutes of that time was measured as I watched the craft hover slowly toward the direction of Phoenix 140 miles to the south....

Jenny Randles documents one account that occurred in Oxford, Maine, in 1975, involving two young men who were in a wooded area near a lake. It was around 3 a.m., and they heard a strange noise. They got in their car to see what could be causing the noise when suddenly the entire car was enveloped in a very strange "colored glow." They were just as suddenly transported a mile away with the car pointing in the opposite direction! Panic-stricken and confused, they were surrounded by a gray fog, and were unable to start the engine. The car had stalled.

In another report from 1995, four people in Calder Valley in England were having a barbecue when they began to experience what they perceived to be "heaviness" in the atmosphere. Everything suddenly felt strange, and they could feel an electrical charge in the air, similar to those at the onset of a thunderstorm. They experienced collective time distortions, as if events were being "compressed" in their flow, and objects began moving of their own volition. All four felt disoriented, and reported the appearance of a dark gray mass and that a mist or fog enveloped the garden area they were in. They recalled as well a beam of light, and then instant dark, "as if many hours had passed in an instant."

Witnesses to these time slips and storms often have a difficult time recalling details of their experience and are frequently so disoriented that Randles suggests that that they might be instantly "jumping reality tracks" and seeing the world shift into a slightly different version than the one they regularly inhabit. She calls them a sort of "reality blinks" and believes that they may happen to us all the time, just not this powerfully and profoundly (again, déjà vu comes to mind!).

These time storms have elements of what may be natural and environmental causes, as indicated by the electrical effects and changes in pressure. But it's the spatial and temporal distortions that people who experience time storms, slips, and shifts never forget. That sense of disappearing into another time and place and emerging again back into "normal" reality suggests that, just beneath the surface of our three-dimensional existence (with time of course as the fourth dimension), there are other times and places waiting to be discovered and explored.

The Most Famous Time Slip

One of the most famous time slips involves two English women, Charlotte Anne Moberly and Eleanor Jordain, the principal and vice principal of St. Hugh's College in Oxford, England, respectively, who allegedly had a time slip in 1901 in the gardens of Versailles, France. The two women were apparently looking for the path that led to the Petit Trianon, the private palace of Marie Antoinette, when they found themselves off that path and on another. They both felt an overwhelming sense of "oppression and dreariness" as they spotted an old, deserted farmhouse and some abandoned farm equipment. They continued walking and encountered men dressed very dignified, who they assumed were palace gardeners. They also passed a cottage where a woman and a young girl stood "frozen in time" and reported that "everything looked unnatural, therefore unpleasant; even the trees seemed to become flat and lifeless, like wood worked in tapestry. There were no effects of light and shade, and no wind stirred the trees." They eventually came across a man that they felt was very foreboding and had an expression of evil (some suggest this was the Comte De Vaudreuil, an enemy of Antoinette's), before encountering another tall man with dark eyes who took them to the Petit Trianon. These visions of encountering strangely and elegantly dressed men and women wearing clothing from another time and place continued, along with the sensation of things not being right and a feeling of oppressiveness, until the women rounded the corner of one building and felt the darkness and oppressiveness "lift" and return to normalcy. So affected were they that the two women wrote a book about it later: *An Adventure*, under the pseudonyms Miss Morison and Miss Lamont.

Whatever happened to these two women, and we can say they were making it all up or dreaming or just misinterpreting normal events, they were deeply affected by the sense of entering an alternate reality and experiencing a much earlier time in history. Time slips, shifts, and distortions happen to people who don't often have the courage to report them, and we can only imagine the true numbers of those who get a glimpse into the other levels of this Grid we call reality.

Imagine, instead of things disappearing into time slips, things coming out of time slips and manifesting here in the now.

Objects Out of Time
By Sally Richards

Figure 8-1: A séance during the 1800s where a key is believed to be apporting for the woman at the table.

Apport. Before I became a researcher of Spiritualism and had a bookcase of pre-1900 books on scientific research of mediumship, I'd never heard the word. Apporting is the ability of an object to not only break the physical barrier of transporting through time and space, but to also reassemble its mass and present itself at the beckoning of a medium.

Although there are belief systems other than Spiritualism that describe such abilities, one of the most near and dear to my heart, being raised in Hawaii, are the ancient kahunas of the islands who claimed to have many supernatural powers.

Normally, in the case of apporting during a Spiritualist séance, a circle of people would sit down and hold hands as the medium, or control, summoned an object—and it would appear from thin air. Although Samantha Stephens (*Bewitched*) made the whole business look quite easy, it was not uncommon for mediums to sit for an hour to bring forth an object; other times it was instant. And the objects seemed nearly always to be a surprise. Physical mediums are now a very rare breed that often suffered from the physical maladies created by the physical energy needed to apport objects and create other physical phenomena.

Today, apporting is mostly forgotten word, and skill. Some skeptics say the lack of electricity and media equipment in the old days allowed charlatans to get away with bilking clients into thinking they had received apported items that materialized from thin air.

I interviewed a woman who is not a medium by profession, but rather by birth. Her family, some with the skill of mediumship, and their friends, some also mediums, would have séances at their home called home circles. One day she sat in on a séance with a physical medium conducting the circle. Often at these séances were long, tin horns—the kind the suffragettes adopted (Susan B. Anthony, who tested for her mediumship credentials at Lily Dale, New York, served on the platform, and Victoria Woodhull (1838–1927), the first woman to run for president, was the personal medium and financial advisor for Commodore Cornelius *Vanderbilt*) to blow in the streets as they marched. This particular séance did have a tin horn on the table.

The woman said she was 12 years old at the time the séance took place, and it was not very dark in the room. She said the horn floated above the center of the table—mouth side down (which was no larger than a half an inch). She said during these circles that sometimes spirit voices would come through the horns, or a gift would appear from them, as they floated toward each of the people sitting. On this particular day, the tin horn pointed to her and she heard a faraway sound like a marble circling the inside of a tin drain on its way down, making scraping circles with each rotation. She said it sounded like it was coming from a mile away. She put out cupped hands in anticipation of whatever was coming down the horn. When the noise stopped, a small gemstone, which she still has some 50 years later, fell into her hands. Others seated at the table also received gifts from the horn, some so large it was difficult how they could have fit through the hole.

There were a good many mediums founded as frauds by researchers and lawmen while apporting. Skeptical investigators found the apported items that were planted in the pitch-dark room, or hidden on the medium's body. Yet, other mediums were tested by strictly scientific methods by investigators and found valid, but without explanation. During these sittings, small animals, insects, plants, and valuable and strange objects appeared—and some disappeared as immediately as they'd appeared. Some believed the reason why some disappeared was because that they were from another time and needed to be returned to keep the quantum order.

Talk of apports (the objects themselves) was all the news back in the late 1800s through to the 1930s. Sir Conan Arthur Doyle (author of the Sherlock Holmes books, paranormal investigator, and later a leader in the Spiritualist religion) investigated mediums all over the globe with friend and fellow researcher Harry Houdini. Researcher Harry Price (1881–1948), a British researcher, also searched for answers. He was giving a talk at the Psykisk Udstilling (in English, the Psychic Exhibition) held in Copenhagen, in the Metz Tea Rooms in the Østergade January 10–18, 1925, when an interesting connection was made that would greatly influence the growth of Spiritualism in the UK. Originally, the talk was supposed to be given by Doyle, but he cancelled and asked Price to speak in his place at the last minute, and Price agreed to do so.

One of the exhibits during the conference was provided by a man named J.S. Jensen, president of the Danish Psykisk Oplysnings Forening (Society for the Promotion of Psychic Knowledge). He had, for years, been collecting thousands of apports. They had a conversation, and Price agreed that the collection should come to London.

In Price's own words:

> It had been arranged that the London Spiritualist Alliance should hold a bazaar and fête on May 20 and 21, 1925, at the Caxton Hall, in order to increase their funds, and I thought that it would provide a good opportunity for staging Mr. Jensen's

psychic relics. This was agreed to, and arrangements were made for exhibiting other objects of interest that owner of such things was invited to send in. The Exhibition was an outstanding success. Crowds from the Metropolis and the provinces flocked to see the many thousands of exhibits illustrating the history, literature, and development of psychical research and spiritualism. I do not know how many people visited the Exhibition during the two days its doors were open, but the rooms were uncomfortably crowded most of the time, and nearly £1000 changed hands during the period of the show. Pressmen who came to scoff were spellbound at the—apparent—evidential nature of some of the exhibits, and all agreed that there is in modern psychical research a strong *primâ facie* case for serious scientific investigation.

Mr. Jensen sent some thousands of objects, mounted on boards that were hung on what are known as "Spanish walls": large wooden frames, connected by angle-irons so that they can be adapted to rooms of all sizes. The Exhibition filled one hall, two rooms, and a gallery—5,000 square feet of exhibition space. Brief descriptions of the articles filled a 36-page Catalogue, that must be unique of its kind, and which will be of some historical interest in years to come.

In a foreword to the Catalogue I warned visitors that no guarantee could be given that every exhibit was what it purported to be, and I am sure that this point was appreciated. Fraud, folly, and self-deception were writ large on some of the articles sent in by their credulous owners, and it was quite pathetic to listen to the marvelous "history" of some of the objects. There were several curious happenings even in the precincts of the Exhibition itself. I was walking up the main stairs with a gentleman—who, for years, claimed that he was a sort of target for a perfect bombardment of "apports" (objects spontaneously precipitated into the presence of a medium by paranormal means), that showered upon him at all hours of the day and night, when he gave a start and was struck on the shoe by a safety-pin "apport." I picked it up and it felt warm. Apparently it had dropped from the skies!

This gentleman was exhibiting his collection of apports, and they filled a couple of suitcases. They included golf balls and cigarette cases, and an African native's leather apron that had been forwarded from nowhere, "by easy stages," he said, especially for this Exhibition! This same gentleman was once riding in a bus full of people when a French milliner's highly coloured hatbox, complete with the latest creation from the Rue de la Paix, was "apported" on to his lap! Unfortunately, the *midinette* from whom, presumably, the hatbox had been "lifted," did *not* appear. This same man had a similar experience in another bus. Suddenly, a large—and hot—coffee urn materialized on his lap.

In 1904, it was psychical researcher named Ernesto Bozzano, one of the most famed Italian psychical researchers, reported a case which illustrates one of the strangest cases of apporting I've heard.

During a sitting in the house of Cavaliere Peretti, in which the medium was an intimate friend of ours, gifted with remarkable physical medium-ship, and with whom apports could be obtained at command, I begged the communicating spirit to bring me a small block of pyrites which was lying on my writing table over a mile away. The spirit replied (through the mouth of the entranced medium) that the power was almost exhausted, but that all the same he would make the attempt. Soon after the medium sustained the usual spasmodic twitchings [that] signified the arrival of an apport, but without hearing the fall of any object on the table or on the floor. We asked for an explanation from the spirit-operator, who informed us that although he had managed to disintegrate a portion of the object desired, and had brought it into the room, there was not enough power for him to be able to reintegrate it. He added, "Light the light." We did so, and found, to our great surprise, that the table, the clothes and hair of the sitters, as well as the furniture and carpet of the room, were covered with the thinnest layer of brilliant impalpable pyrites. When I returned home after the sitting I found the little block of pyrites lying on my writing table. Missing from it was a large fragment, about one-third of the whole piece, which had been scooped out.

If one chooses to believe this story, considering the time in between the request and the outcome, time travel certainly has to be considered as having been used. One also had to have a way to get into Ernesto Bozzano's home, which he claimed had been secure. In my years in researching the physical aspect of mediumship and the materialism of apports, there always seems to be an element of manipulating time physically. So why does this continue to be a theme in the interviews I've conducted and my research of the archives I've collected of books regarding physical research of mediums from 1850 [to] 1930?

I imagine what it means, amazing as it may seem, is that some people throughout time have had the ability to manipulate time and space and transport items with mass from one place to another without touching them. You don't hear much about this kind of thing anymore unless it's someone like David Copperfield, a magician, making an elephant appear out of thin air.

I had a personal experience a few years ago when I was on my way to a Meetup group dealing with psychic abilities. It was in the University Heights, a hip San Diego neighborhood, when a man approached me and handed to me a card and said, "You'll be needing this." I nodded, smiled at him, and put the card in my pocket, thinking it was some kind of solicitation. I thought he looked a little too well dressed for the role of street hawker, and there was something about his look of intense conviction when he spoke that gave me reason to pause. I turned around to give him a second look, and he was nowhere to be seen.

I took a few steps back and looked in the coffee shop I'd just walked past and he wasn't there, and even if he'd burst into a run it would have been difficult to get out of my eyeshot that quickly. I pulled the card out, it had his name—John something—and his title: Time Traveler. The card had a weird logo on it. I shoved the card back into my pocket and went to find the meeting that I was already late for. Later, when I tried to find the card again, it had just disappeared. Although I searched everywhere, I couldn't find the card and assumed it had fallen out of my pocket. A few months later I came upon the card on a bookshelf in my library; I placed it on my desk and was ready to call the number the next day to find out what it was all about. The next day when I went to find it, it had disappeared again—and not resurfaced in two years.

Sally Richards is a Spiritualist medium and a researcher of historic mediumship. Her book Historic Lily Dale *will be released in January 2013, and* Ghosthunting in Southern California *(Clerisy Press) will published in 2012. She has one of the largest private collections of Spiritualist artifacts from the Antebellum and Victorian eras and 1930s, including unpublished manuscripts from philosophers such as Manly Hall and post-mortem photographs and mourning jewelry.*

Another fascinating time anomaly involves the world of time shifts. We turned to an expert on the subject for a complete description of what time shifts are and how they operate.

Time Shifts and Starfire Tor

Starfire Tor (*www.starfiretor.com*) is a respected time anomaly and paranormal phenomena expert who discovered the Core Matrix, Time Shifts, Co-Existing Time Lines, and the Unified Field Theory of Psi. Starfire Tor is also a scholar, writer, producer, lecturer, composer, and television and radio personality whose acclaimed research also includes reality shifts and time travel. She is the CEO of Starfire Communications and the founder of the Whale and Dolphin People Project. We asked her to answer some questions about her ongoing research.

Jones and Flaxman: *How did you become involved in this research?*

Starfire Tor: From the time that I was a child I experienced many different types of paranormal, psychic, and time anomaly phenomena. This included accurate precognition of future events, some of which ended up saving lives. Even as a child I sought out the answers as to what was behind the abundance of authentic phenomena I was experiencing. But there were no tangible scientific answers to the phenomena from either mainstream science, advanced cutting-edge science, fringe science, or from people promoting spiritual and new age beliefs.

And so it was, as a child, that I began my quest to uncover and understand the core science that would explain how all of these paranormal, psychic, precognition, and time anomaly experiences were possible. Over the years the depth and complexity of my research grew, and I discovered that the type of experiences I was having was not unique. Both historically, and present day, millions of people have had similar authentic experiences. This made my research mission even more important because it wasn't just about solving a multitude of personal paranormal and anomalous events. My quest had evolved into a search to uncover the all encompassing, cohesive, and scientifically incorruptible unified field theory that would explain the source and mechanism of all of the events. It was on this quest that I discovered the Core Matrix, Time Shifts, Co-Existing Time Lines, and the Unified Field Theory of Psi.

Jones and Flaxman: *You discovered the Core Matrix, Time Shifts, and Co-Existing Time Lines. Can you explain them in simple terms? How do time anomalies and time travel fit in? Are Co-Existing Time Lines and parallel time lines the same thing? Are Time Shifts and Reality Shifts the same thing?*

Starfire Tor: The Core Matrix, Time Shifts, and Co-Existing Time Lines are the greatest secret that no one knew existed until I discovered them. By design, they function as integrated information collectives and repositories that create, stream, edit, and conceal the many time lines, realities, and dimensions that make up our true existence. They work in concert to create the illusion that we live in a world where the space-time continuum is linear, and that it flows unchanged from the past to the present and into the future within a single time line. Nothing could be further from the truth.

Using simple analogies: The Core Matrix is like the hard drive in a computer whereby all data, programs, and hardware are stored so that they can be used when needed. This means that all of the software, hardware, elements, and data needed to stream every Co-Existing Time Line are stored in and accessed from the Core Matrix. Co-Existing Time Lines are like multiple saved versions of the same Word document, with each being different because elements in the Word document were changed through editing. The newest revised version of the Word document is likened to the post Time Shift revised new dominant time line. The older versions of the Word document, while not actively being used, are kept intact and in existence by saving them to the hard drive where they are ready to be accessed should any part of an older document be needed. Time Shifts are akin to the keystroke action, which creates the edits in the Word document that makes them different from all other versions of the same document.

Our brains are the interface to the Core Matrix, as well as being the transceiver through which our perception of reality, space, and time are formed. Therefore, it is through our brain and Core Matrix interface that the Time Shift program is able to conceal its time line changing activities. As confounding as this stealth activity is, the same mechanism that makes the activity undetectable to the conscious brain is the same mechanism that can locate and exploit glitches in its stealth program. These glitches create situations in which the brain is not totally assimilated into accepting the new dominant time line as if it is the only time line in existence. The lack of total assimilation creates a situation where the brain retains two conscious memories of the same elements and scenarios, but in two different Co-Existing Time Lines. This situation creates a condition that I call dual Co-Existing Time Line memory conflict. It's a condition that millions of people have experienced. It's also a basic way in which time anomalies, many of which are the result of time line edits, are experienced and remembered.

In a dual Co-Existing Time Line memory conflict, which is a marker of Time Shift time line editing activity, a person can clearly remember two events that should not exist in the same time line at the same time. There are many variations of this, with the most encountered one being something that I call the Time Shift Living Dead Phenomena. The TSLD phenomena are identified as such when someone has a clear memory that someone who they absolutely know has died is suddenly alive again. The reverse can be true as well, when a person recalled as being currently very much alive is suddenly discovered to have died a long time ago. My "The Time Shift Living Dead

Phenomena" article, available on my Website, explains more about the experience and the science behind it.

Sometimes dual Co-Existing Time Line memory conflicts stand alone without physical documentation, while other times the Time Shift glitch provides both a dual Co-Existing Time Line memory conflict and the physical evidence that backs it up. My big breakthrough, in confirming the validity of my discovery of Co-Existing Time Lines, came when I was able to acquire the physical documentation that proved that more than one time line was at play. Since then I've been able to gather many pieces of physical evidence proving the existence of Co-Existing Time Lines. One involves irrefutable multiple-source documentation showing that the same person was in two different places at the same time thousands of miles apart. The same type of documentation, with one being a mainstream large circulation newspaper reporting the unique event, showed the same unique event happening in two distinctly different years.

I coined the phrase *Co-Existing Time Lines* to create a more accurate understanding of the existence and creation of multiple time lines. Some time travel and quantum mechanics theorists, as well as a number of science fiction books, films, and television shows, have promoted the idea that there are parallel time lines. By parallel, that would mean existing side by side like railroad tracks that run next to each other. What I discovered is that time lines do not exist as parallel items. Instead, all time lines co-exist in the same space and are separated only by their individual frequencies. I coined the phrase *Time Shift* to describe the action and outcome in which our conscious reality, which our brain is programmed to perceive as the one and only time line existing and experienced, is constantly being physically and temporally edited and restructured. All Co-Existing Time Lines and unstructured time line elements, whether consciously realized or not, continue to stream from the Core Matrix. The brain constantly accesses this streaming data while awake and asleep.

Jones and Flaxman: *Are Time Shifts and Reality Shifts the same thing?*

Starfire Tor: As I explained, our brains interface with the Core Matrix. Because of this, we're plugged into, and entangled with, the Time Shift mechanism and elements streaming in Co-Existing Times. Yet whether singularly or collectively, humans cannot trigger Time Shifts, nor cause time line edits that create new time lines. What humans can do is create a Reality Shift, which is a natural human ability whether people are consciously aware of it or not. It works by copying desired elements from Co-Existing Time Lines and pasting them into the current dominant time line. Because it's a copy-paste technique, much like the copy-paste feature on a computer, Reality Shifts do not destroy the elements that they copy, and the manifestation itself does not create a new time line. Time Shifts can create Reality Shifts, but Reality Shifts cannot create Time Shifts. Reality Shifts are a natural part of being human, whether a person is consciously aware of it or not. While I did not discover the existence of Reality Shifts, I did solve the mystery of what they are and how they work. Based on this I created the psience protocols I named Reality Shift Manifestation and Reality Shift Healing, which are both personal and group help tools. I also created the global collective Reality Shift based protocol the Vortex Peace Prayer.

Jones and Flaxman: *How does a Time Shift affect us in a global and a personal sense?*

Starfire Tor: Due to the brain's interface with the Core Matrix, humans are entangled with the Time Shift process. However, whether singularly or all together, humans cannot trigger or control the activity of Time Shifts and the time line edits that they bring. One of the power sources that trigger a Time Shift is a specific type of solar activity that affects Earth's magnetosphere, such as coronal mass ejections, solar flares, and solar winds. People may not be able to trigger and control Time Shifts, but it doesn't mean we're helpless. Because of the brain's interface with the Core Matrix and Co-Existing Time Lines, our ability to access and utilize meaningful information and skills is near limitless. Anyone can choose to lift the veil of illusion and evolve into a person, and a united people, whose relationship with the cosmos and each other is free of disinformation and the pettiness that keeps us shackled to the illusion. But to do this you must learn about our authentic existence and how it really works, and that means learning about the Core Matrix, Time Shifts, Co-Existing Time Lines, and the Unified Field Theory of Psi....

Another perk of Time Shifts is that there's no need to wonder if we survive 2012. We do. Time Shifts involve time looping, and time looping involves a repeat of time line elements. Therefore, we've been through versions of 2012 a number of times, and we're still here regardless of what dangers may come.

Time Shifts don't just edit time lines. As a byproduct, the mechanics of time line editing can trigger devastating Earth changes. Due to frequency changes, associated with Time Shifts and the nature of Co-Existing Time Lines, the time line restructuring process can result in the imperfect synchronizing and alignment of the edited elements that make up a time line. Earth has its own global electromagnetic resonance, which when disturbed can cause releases of uncontrolled energy. When a Time Shift process results in imperfect editing synchronization and alignment, the imperfection creates space-time and physical time line disturbances that can trigger earthquakes, tsunamis, volcano blows, geophysical stresses, unusual and dangerous weather, and even whale, dolphin, and porpoise stranding.

Two examples of Time Shift entangled global disasters are the September 11, 2001, attacks in the United States, and the March 11, 2011, earthquake and tsunami in Japan. Both of these deadly events occurred during a Time Shift Alert, which was activated in part by solar activity hitting Earth's magnet field. Before each of these disasters struck I posted warnings specific to those events. FYI: Solar activity does not directly cause earthquakes. Some solar activity triggers Time Shifts, some Time Shifts trigger earthquakes, but not all earthquakes are triggered by Time Shifts.

Due to the mechanism involved in time lined editing, Time Shift activity can create and amplify psychic, paranormal, and precognitive experiences. Besides the Time Shift information that I was releasing prior to September 11, 2001, shortly before the attacks I had a series of vivid and upsetting precognitive psi data downloads. On September 10, 2001, I shared a compilation of them online. I wrote about airplanes and danger to the World Trade Center, danger to U.S. military buildings, a coming war, devastation in the finance community, and more. I also sold my stocks. On September 11, 2001, the attacks killed thousands and changed the world forever. Practically unremembered on September 11, 2001, is Hurricane Erin, the Category 3 storm heading toward New York and the East Coast. Time Shift time line edit, and it changed course, leaving clear blue skies over New York and New England. Had Hurricane Erin hit New York, the weather would have prevented the devastation at the World Trade Center, the Pentagon, the Shanksville crash, and maybe stopped it from happening on any day. My September 11, 2001, premonition can be accessed in my Starfire Tor Future Event Precognition section on my Website.

For years I've been releasing Time Shift Alerts. Part of the criteria involves the emergence of the type of solar activity I've already described. There are other criteria. At the same time that I release a Time Shift Alert I release a whale, dolphin, and porpoise stranding alert, although in different arenas. This is because the same solar activity that can trigger a Time Shift also causes a disturbance in Earth's magnetic field that can disrupt the bio navigation and communication abilities of cetaceans. This can cause them to strand and engage in wrong-way travel. When it came to the March 11, 2001, Japan disaster, elements of both provided the information that I needed to release earthquake warnings.

Here is the sequence of events:

February 24, 2011—I released a Time Shift Alert, in part, because an M3 class solar flare had erupted. The energy was due to hit Earth between March 3 and 4, 2011. Ultraviolet radiation was involved, radio frequency signals were disturbed, and from a separate solar event a solar wind was expected. I also released a cetacean stranding alert.

March 1, 2011—I updated both alerts when an unexpected solar wind stream hit Earth causing a magnetic storm.

March 3, 2011—The originally expected solar wind arrived.

March 4, 2011—A large pod of melon-headed whales stranded on the shores of Japan, while a high-speed solar wind stream buffeted Earth's magnetic field. Since these are not the whale or dolphin species that Japan usually slaughters, which is an unacceptable barbaric activity that must end now, 22 of the stranded whales were rescued.

March 5, 2011—I updated both alerts because a coronal mass ejection would hit Earth on March 6, 2011. I updated the Time Shift Alert to a Time Shift Cluster and extended the stranding alert.

March 7, 2011—An X2 class CME erupted, and is expected to hit Earth causing a geomagnetic storm March 9 or 10, 2011. A review of the data, along with the whale stranding in Japan during the Time Shift Alert, caused me to believe that Japan was in danger of being hit with a massive earthquake during the Time Shift Alert window. I've seen the pattern before.

March 9, 2011—Earth-orbiting satellites detect an X1-class solar flare and that the March 7, 2011, solar flare is traveling faster than any other solar flare since September 2005. The energy from both flares may hit Earth on March 9–10, 2011. I update my Time Shift Alert and Stranding Alert and warn that Japan is in imminent danger of experiencing a massive Time Shift triggered earthquake between March 9 and 12, 2011. I implore the Japanese to stop killing cetaceans and release the dolphins in the Taiji cove before the earthquake hits.

March 10, 2011—I report that the expected CME hit Earth's magnetic field.

March 11, 2011—Earth's magnetic field is still reverberating from that CME strike, and then a magnitude 9.0 undersea mega thrust earthquake erupts off the coast of Japan. It's the most powerful earthquake known to have hit Japan, and triggers a devastating tsunami. The event changed the shape of Japan's land mass and moved it 8 feet. NASA reported that the Earth's axis may have shifted and the length of days may have shortened.

Jones and Flaxman: *We have experienced time slips, which we wrote about in our book* The Déjà vu Enigma. *Do you have a personal time slip experience you can share?*

Starfire Tor: A time slip is the product of a Time Shift. All Time Shift activity causes time slips, but most time slips are never consciously realized or remembered. A Time Shift involves time line editing, the creation of a new dominant time line, and the relegating of the pre Time Shift dominant time line to the status of lesser Co-Existing Time Line. All Co-Existing Time Lines exist in the same space and are only separated by their individual frequencies. Regardless of how many Co-Existing Time Lines there are, or how similar in content many of them may be, no two Co-Existing Time Lines run in perfect sync. Therefore, even the smallest time line edit between two nearly identical Co-Existing Time Lines will cause a jump edit effect. Such jump edit effects are not intended to be consciously realized or recalled, and when they are the reason is usually a glitch in the Time Shift program.

I've investigated many time slip events and the jump edit effect is by far the most amazing and physically undeniable. In one case a jump edit caused people to leave an area hours before they originally entered the area. In another case a jump edit caused a car, driver, and passenger to suddenly be driving on one road, and in the blink of an eye find themselves miles away on a completely different road. In these cases, sometimes missing time is experienced and sometimes time has appeared to go backward.

Previous page—Figure 8-4 and Figure 8-5: The Magic Castle in Hollywood was the location of the time slip investigation with the Striebers. The second image shows the only door in and out of the small ladies room where a woman was seen exiting, but not entering. Starfire Tor was alone inside the small ladies room, Anne Strieber had just exited and was standing outside next to the door, and Whitley Strieber was at the top of the staircase with a straight line of sight to the door when the woman exited. She was never observed entering the windowless room. The investigation concluded that the time slip was caused by a Time Shift time line edit.

I am pleased to say that along with Whitley Strieber, Anne Strieber, and Brandon Scott, I was involved in what turned out to be the best documented and best researched time slip in history. It happened in 2006 when Whitley, Anne, and I went to the Magic Castle in Hollywood, California, to have dinner and catch Brandon's magic act. Between finishing eating and going to the theatre Brandon was performing in, we took a stop in our respective rest rooms. It was during this seemingly ordinary activity that an amazing jump edit time slip took place involving the sudden manifestation of a mystery woman who somehow appeared in a small windowless, one-door room without going through the normal route of walking through the door to enter the room.

The entire incident and report can be read on my Website, which include links to individual special reports by Whitely and Anne. After a thorough investigation, which included tracking down and interviewing the mystery women, accessing blueprints of the room to ensure that there were no secret doorways into the room, and much more, the investigation was conclusive: The mystery woman, Anne, and I had been consciously involved in a time slip that merged elements of two time lines into a single event. Different time line sequences merged in a near collision involving same space occupation, a calamity that was narrowly averted. Watching it all go down, from a bird's-eye view of the only door in and out of the room, was Whitley.

Besides my published investigation, the time slip event was featured on the History Channel, the Discovery Channel, Coast to Coast AM appearances on separate shows with hosts George Noory, Art Bell, and Whitley Strieber, and several appearances on Whitley Strieber's *Dreamland* show and Anne Strieber's *Subscriber Show*.

Jones and Flaxman: *Do your discoveries resolve the grandfather paradox in time travel?*

Starfire Tor: My discovery of Co-Existing Time Lines not only solves the riddle of the grandfather paradox, it removes it entirely from consideration as a valid time travel topic. The grandfather paradox has been a staple of science fiction writers, as well as quantum mechanics theorists, for decades. The paradox sets up an intriguing time travel brainteaser: What would happen if a person travels back in time and kills his own biological grandfather before the grandfather ever met the time traveler's grandmother? This would imply that one of the time traveler's parents would never have been conceived, and by extension the time traveler would never have been conceived. Since the time traveler is alive and did manage to go back in time and kill his biological grandfather, how could such an act be possible?

The grandfather paradox is based on the assumption that there is only a single time line, and that it is within this single time line that a time traveler can move backward and forward in time. My Co-Existing Time Line discovery proved that not only are there an untold number of time lines that co-exist in the same space, separated only by their individual frequencies, but that the very act of time travel causes the time traveler to switch to another Co-Existing Time Line. The true mechanics of time travel make it impossible for anyone or anything to time travel within the time line from which they began their time travel journey. Since no Co-Existing Time Lines, including who inhabits them, are exactly the same it possible for a time traveler to shoot his biological grandfather in a Co-Existing Time Line. Since it would be unknown who this Co-Existing Time Line grandfather might have married and conceived, had he not been killed, it cannot be determined whether the time traveler was ever conceived in the Co-Existing Time Line in which the grandfather was killed.

To learn more about how different Co-Existing Time Lines work, and how the people who inhabit them can be born and die at different times in different Co-Existing Time Lines, read my article "The Time Shift Living Dead Phenomena" on my Website.

Jones and Flaxman: *How does dreaming tie into time anomalies and time travel?*

Starfire Tor: Part of my discoveries, by extension of learning about the Core Matrix, Time Shifts, and Co-Existing Time Lines, was to learn the truth about how and why we dream. You can forget about most of what mainstream dream experts claim about dreams being the brain working out the days problems through some interior process. The truth is, the brain is the interface to the Core Matrix and its wealth of streaming data. Most dreams are an offshoot of the brain interfacing, accessing, and sometimes downloading this streaming data. Because so much of that data is raw, and is not cohesively connected to a well-programmed streaming Co-Existing Time Line, the brain has a hard time creating a logical and linear story from the data fragments. This is why so many dreams appear to be so odd, weird, and fragmented. However, there are people who have brain interfacing capabilities that are more adept at accessing tangible Co-Existing Time Line data. It is at these times that such people can access information that will prove to be precognitive, and even access Co-Existing Time Lines in a manner that constitutes a form of brain interfacing time travel. Accurate precognition, using this mechanism, would be a type of time travel that has the potential to involve physical aspects beyond a brain interface.

I've been blessed to be a person with precognitive skills, which have not only saved lives but have given me insights into the science of the dreaming brain and the many things that it can do. To learn more please visit the Starfire Tor Future Event Precognition section of my Website.

Jones and Flaxman: *There are people who have publically claimed that they were or are either time travelers, part of a time travel program, or associated with something that involves time travel. What do you think about these claims, and how can someone tell if a time travel claim is authentic or a hoax? You did in-depth research into whether the Philadelphia Experiment is fact or fiction, and you did the same for the Aerial Drone mystery. What did you discover, and how can people access your investigations?*

Starfire Tor: Time travel is one of those topics that have captured the imagination of the world for a very long time. It's an appealing, adventurous, and romantic notion whose popularity is reflected in the wealth of time travel themed books, movies, television shows, documentaries, personal stories, and conspiracies. But our attraction to time travel may not be solely rooted in our imaginations alone. Our attraction to time travel may actually be rooted in a Core Matrix memory of Co-Existing Time Lines and time travel itself. Consider this: Based on my own research, which has demonstrated that our brains are the interface with Co-Existing Time Line data that streams from the Core Matrix, I believe that at some point in some Co-Existing Time Line, time travel has already been developed and utilized. Therefore, it is logical to consider that time travel already exists, and that we have already encountered it somewhere in time.

Because the notion of time travel is so deeply rooted in our conscious and unconscious brain, a time travel hoaxer can find a ready-made audience willing to believe the most illogical and untruthful of claims. As the discoverer of Time Shifts, the Core Matrix, Co-Existing Time Lines, and the Unified Field Theory of Psi, I am in a unique position to be able to determine when a time travel claim is sincere, when a time travel claim is an intentional hoax, and when a time travel claim is a lie that stems from the delusional mind of the person making the claim. Time travel hoaxes, for whatever dishonest reason that they are perpetrated, are nothing more than attention-seeking unsponsored disinformation that undermines the integrity of authentic public time travel discourse. Although time travel hoaxes do not negatively impact authentic time travel-based projects, the disinformation can attract and misdirect the energy and belief systems of well-intentioned but gullible people. Since hoaxers are without ethics, they have no qualms about tricking susceptible people into promoting their lies and personal agendas.

Therefore, having tools of discernment are about more than protecting the integrity of the subject matter. They are about protecting honest well-meaning people from being used and abused by hoaxers and their self serving personal agenda. I could easily generate a list of hoaxers by name. Instead I'm going to share some of the master keys of discernment, which consist of a list of warning signs that you need to watch for with every time travel claim.

For the most part, time travel hoaxes contain claims that are so illogical, and counter-intuitive, that you would have to throw out your common sense to believe them. But not all time travel hoaxes are so obvious, with the Philadelphia Experiment being at the top of that list. Like so many, I was attracted to the basic story because there was a sense of plausibility about it. Not the stories of those who latched onto the myth, but the myth itself. I wanted to know the truth because the myth was so interwoven into time travel belief systems, and it permeated the time travel researcher and enthusiast community. I tracked down the truth and broke the story many years ago. The Philadelphia Experiment never happened. My research revealed that what appeared as plausible was actually the illusion of plausibility.

I recommend reading my entire investigation to get the full impact of what I uncovered, including documents and photographs, "The Philadelphia Experiment Hoax Report," available on my Website.

The Philadelphia Experiment: In brief, it was claimed that in 1943 the USS Eldridge DE-173 was involved in an experiment to render it invisible. In a U.S. port it was rigged with an invisibility device. When the device was activated the boat became unstable and went phasing through space and time to various locations. A version of the myth claims that the boat time traveled, and so did some of the crew. When it was finally retrieved some of the crew were molecularly embedded into the boat's structure, and some later burst into flame.

With common sense, my research, and information about U.S. military procedures you don't have to have an advanced degree in science to understand that it's a hoax. If the Eldridge had been involved in an experiment as claimed, and given the outcome as claimed, the boat would have been pulled from military service, it would have been relocated to a secure site where it could be contained and studied, and under no circumstances would it ever be allowed to leave U.S. custody. But that's not what happened. My research proved that the USS Eldridge DE-173 did serve after the claimed experiment, was eventually decommissioned, and was sold to Greece, where per regulations it was given a new name and number. Common sense dictates that with this real-world history the Eldridge could not have possibly been involved in the experiment as claimed. Conclusion: Hoax.

The Aerial Drone mystery, which gained momentum through the internet, is also a hoax. I recommend reading my entire investigation to get the full impact of what I uncovered, "The Truth About the Aerial Drone," including documents and photographs, on my Website.

The aerial drones came on the scene in 2007, and with pushes from several directions it went viral. Both physical and digital images were involved, as was detailed nonsense data claiming ET reverse engineering, a claimed alien alphabet, possible time or dimensional travel, and alleged witnesses. Some people did their best to seriously investigate the mystery, while others seemed to be more interested in promoting themselves through the mystery rather than seriously investigate the mystery. I put in time and brains and eventually broke the case wide open.

Here's the bottom line: I tracked the hoax down to a Dell Computer owned Alienware promotion for their Area-51 laptops. The contest was advertised as "Decipher the Alien Message and Win a Trip to New York City." It featured the same character symbols as those found on the aerial drone and anti-gravity generator hoax items. Some of the alien letters were even stamped into the computer cases. Conclusion: Hoax.

Unfortunately, a short-lived television series called *The Sarah Connor Chronicles*, which was a spinoff from the time travel-themed *Terminator* movies, inserted similar-looking aerial drones into a plot line. It just confused people all over again.

I could list other time travel hoaxes, but I've rather give people the tools to discern what disinformation is and what true information is. The first step is to read my Philadelphia Experiment and Aerial Drone hoax investigations. It's not just about the conclusions. It's about the methodology I applied that solved the mysteries. And keep this in mind: Like all knowledge of value, some of the key discernments require research while other key discernments require little more than applying common sense. Some of the truest experiences can contain the most far reaching of concepts. Yet even the most truthful, and far-reaching, experiences contain a thread of recognizable common sense that is tangible and relatable. Here are some of the top rules for ferreting out time travel-related hoaxes and disinformation.

1. If someone claims that the aerial drones and related claims are real, even that they are an insider or a witness, you now know that is not true.

2. If someone claims to have been a part of the Philadelphia Experiment, you now know that is not true.

3. If someone claims that a project they were part of was an extension of, or associated with the Philadelphia Experiment, you now know that it is not true.

4. If someone tries to get around the fact that the Philadelphia Experiment is a hoax, by claiming that the USS Eldridge wasn't the real boat involved, that's either a person not willing to face the truth or someone trying to salvage their own game of disinformation.

5. If someone claims to be a time travel authority and researcher, and they endorse and promote the Philadelphia Experiment or the Aerial Drones, then this is not someone who is able to discern which time travel info is real and which is untrue.

6. If someone's time travel claim contains story elements from a time travel-themed science fiction film, television show, or book, then the chances are very high that this is the actual origin of the claim. Most time travel-themed entertainment vehicles do not have story lines that incorporate authentic information on time travel science—no matter how good the shows are. This includes *Back to the Future, Terminator, Quantum Leap, Somewhere in Time, The Time Machine, Stargate: SG1,* and some *Twilight Zone* episodes.

7. If elements from someone's time travel story have changed, or been omitted, the chances are very high that the changes were made to create a more believable story with a more believable scientific view of time travel. Watch out for time travel claims where time travel adventures consist of traveling forward and/or backward within a single time line. The act of time travel involves shifting to Co-Existing Time Lines, because it's not possible to time travel within the Co-Existing time line from which you began your time travel journey. Hoaxers who change their single time line story, by newly incorporating multiple time lines, can be found out.

8. If elements from someone's time travel story includes specific information about his or her past, which is claimed to be our near future, and upon reaching the dates none of the information comes to pass, then chances are very high that the time travel claims are not true.

9. If a person claims to have been part of a U.S. government time travel program, and has gone public with the story, chances are very high that this person has no such history. It must be considered that, should such a program exist, the caliber and character of the participants would feel honor and duty-bound to keep the team secret. A time travel mission would be too globally important and delicate an asset to risk with anyone of less

commitment. Also, breaking the silence would expose a participant to fines and prison.

10. If a time travel claim involves publicly dropping the names of powerful people or organizations and daring them to come forward, chances are very high that the person is engaged in a calculated publicity stunt. The person probably knows that no one powerful will allow themselves to be used in a publicity stunt. So the person can continue their game without the worry of immediate consequences.

11. If a time travel story-teller connects his or her story to a known hoaxer's time travel story, then chances are extremely high that the secondary time travel story is also a hoax. Just because more than one person claim's to be part of the same time travel experience does not mean that the claim is true.

12. If someone claims that it's possible to jump from one time line to another, to better your life or for whatever reason, don't believe it. That is not the true mechanism of how Co-Existing Time Lines or Reality Shifts work, nor is it the interfaced relationship that we have with the Core Matrix. Some of the people making this claim are hoaxers whose agenda is to get your money into their pocket. Others are genuinely sincere in their belief, and have no plans to take any money for sharing what they believe to be true. Whether a fraud or sincere, both are spreading disinformation.

For more information on Starfire Tor's work, visit her Website.

The mind and imagination have always been able to travel where the body can't, but in these reports and experiences, it may very well be the body can often go along for the "trip" as well. Time travel, then, may be happening to us all the time, and yet we are so focused on the scientific angle—the provable and objective riding along in a device that takes us back to the past or forward into the future—that we forget we are already journeying. This may be as close as we ever get to real time travel, but maybe it's close enough. Ask those who have experienced mental time travel, precognition, out-of-body and lucid dream travel, time distortions, time storms, time slips, and time line shifts, and they will tell you that what they experienced was as real as anything they've experienced in their normal day-to-day reality. No Tardis or Delorean required!

CONCLUSION:

ONLY TIME WILL TELL

I never think of the future—it comes soon enough.
—Albert Einstein

The future influences the present just as much as the past.
—Friedrich Nietzsche

The future ain't what it used to be.
—Yogi Berra

Undoubtedly, by the time this book hits the shelves a lot will have changed. Some of it will still look the same. Some of it will be completely new and novel to us. That is the nature of the passage of time.

By the time you read this book, we may have breached light speed, mastered the manipulation of the clock, and somehow figured out a way to cram 50 hours into a 24-hour day. That is the nature of scientific advancement.

By the time this book is in your hands, you might be holding a device to read it with that is smaller, faster, and cheaper than the one you are reading books on as we write this. That is the nature of progress.

Time travel was once only the imaginings of novelists and film-makers, story-tellers who looked far into the future through a looking glass called the imagination, with little regard for the ground their stories were breaking in "real world" science. That groundbreaking has occurred rather fast, if you think about it, exponentially increasing in scope as time passes and technology advances. Of course, good ol' human ingenuity races alongside, always keeping pace, even if we do get out of breath a bit here or there.

We may be on the verge of a major discovery within the next few years. Maybe we will discover a way to surpass the speed of light that will allow for more than just a tantalizing glimpse into the possibility of time travel. Perhaps we shall discern proof of the Multiverse at the LHC. And maybe—just maybe—we will discover a verifiable and work-able time machine in the basement of some maverick hotshot professor who as of this writing is anonymous—some invisible genius, working diligently toward an Omega Point when the impossible becomes not only possible, but almost trivial and mundane.

Writing about time travel is all about breathless imaginings and stunning possibilities. But it is also about other kinds of "what ifs" such as those involving the ethical question of "If we can, should we?" If we can time travel someday soon, should we do it, knowing the possible outcomes involving the paradoxes, the Butterfly Effect, and cause and effect? Unless we really do find proof of parallel universes and how to access them, and thereby time travel without paradoxical problems, we are stuck with what we have, and what we have calls us to question whether or not we have the right to change the past or alter the future.

In an interview for *Parade* magazine in September 2010, Stephen Hawking was asked about time travel and responded by stating his current belief: that it's possible, but not feasible, and would require knowledge of things we don't yet have, such as how to warp space-time and what kind of negative energy density matter would be involved. Hawking did bring up a very important point to remember though. He likened us to fish in a fishbowl, in need of a greater "cosmic" per-spective. Like the fish looking out of his bowl, and seeing only one universe—his version of it—we might wonder if we ourselves are in some kind of giant fish bowl, and our perspective of reality be our version of it and not necessarily the entirety of reality itself. In other words, we may have no clue as to what the truth about our own uni-verse is. On a fun side note, Hawking was also asked to provide his for-mula for time travel to the editors of *The Face Magazine* back in 1995 for their special 15th-anniversary issue. His response was a very brief, suc-cinct fax sent to the editors: "Thank you for your fax. I do not have any equations for time travel. If I had, I would win the National Lottery every week. —S.W. Hawking."

Clifford Pickover, writing for *NOVA* in the October 22, 1999, "Traveling Through Time" put it so wonderfully by stating:

Don't believe anyone who tells you that humans will never have efficient technology for backward and forward time travel. Accurately predicting future technology is nearly impossible, and history is filled with underestimates of technology:

"Heavier-than-air flying machines are impossible." (Lord Kelvin, president, Royal Society, 1895)

"I think there is a world market for maybe five computers." (Thomas Watson, chairman of IBM, 1943)

"There is no reason for any individual to have a computer in their home." (Ken Olsen, president, chairman and founder of Digital Equipment Corp., 1977)

"The telephone has too many shortcomings to be seriously considered as a means of communication. The device is inherently of no value to us." (Western Union internal memo, 1876)

"Professor Goddard does not know the relation between action and reaction and the need to have something better than a vacuum against which to react. He seems to lack the basic knowledge ladled out daily in high schools." (*New York Times* editorial about Robert Goddard's revolutionary rocket work, 1921)

"Who the hell wants to hear actors talk?" (Harry M. Warner, Warner Brothers, 1927)

"Everything that can be invented has been invented." (Charles H. Duell, commissioner, US Office of Patents, 1899)

So let's play with this philosophical dilemma a bit and try a "what if" of our own here. As the authors, we, Marie and Larry, propose that we send a fictional character named Freddy Freeforall back into the past from today's date. Having written this book, we now understand the possibility that even the tiniest of changes to the past might have immense implications for the present and the future—but we are snarky and want to test that theory. So we decide to send Freddy back anyway.

But before we do so, we feel it is important to first give you a bit of background on Freddy. He is a holdover hippie from the 60s who went on to create a huge hi-tech empire that employed millions of people and gained him billionaire status. He is married to Jill Goodnplenty, and they have two 8-year-old boys, Rogerthis and Rogerthat (twins, of course). Perhaps you might find this a bit too unbelievable? It's our character, and we can do what we want with him.

Freddy has lived a fairly good life, with few major challenges. His parents are still alive and healthy, and he lives in a gorgeous mansion on a private island with its own sub-oceanic tunnel to the mainland. He is now contemplating running for political office as a progressive libertarian capitalist green. It's a new party, after all. (Like we said, it's our character.)

So without having to fill in every minor detail of Freddy's life, let's just say we send him back to a single day when he is 10 years old. This day was important to Freddy because it was the day he lost his team's Little League game by dropping a fly ball. On this seminal day, Freddy went home, crying, to his beloved computer, where he found solace writing code and creating Websites for people for several hundred bucks a pop (his parents owned the account in their name; it was all legal—sort of). That game, and Freddy's failure, was a major turning point in his life, because it proved to him that he was not athletic, and that he should and must wave his geek-freak flag high. It was a major cause that led to a series of effects that would eventually lead him to his presently awesome life.

What if we sent Freddy back—and we made him catch the ball?

Would he still go on to find solace in his computer, meet another tech geek girl, and marry her many years later? Would Freddy go on to have twin boys and launch a geek-tech empire, happily employing people who went on to have great lives because of the financial security Freddy provided them through stock ownership, profit sharing, and all-around feel-goodery?

Or would Freddy have thought, perhaps—just perhaps—he was an athlete after all and kept at Little League, despite the fact that he would always be mediocre at best? But he had that one great winning catch, and so he ended up going out for sports in college, marrying a cheer-leader who later cheated on him with the football team captain, never had children because he was emotionally devastated for years, and hid himself in his basement using his computer to create a manifesto that would get him into a lot of trouble with the FBI a few years later. And maybe, even worse, he would evade the FBI, take revenge on the cheerleader, and end up in jail miserable and bereft. Heck, he could even have ended up dead in a shoot-out with the Federales.

And all because he caught a damn ball when clearly he shouldn't have.

It's a silly example, and maybe a bit extreme, but think about it. One small change *changes everything*, including the lives of other people: Freddy's children were never born. His wife may have ended up marrying a horrible South American dictator or, worse, ended up being an old spinster working at a library—oh wait, that was *It's a Wonderful Life*. And his company that happily employed so many people who in this economy cherished their jobs? Never existed. By the way, all those people now are on welfare because they could never find stable work.

In that particular time line, Freddy caught the ball and messed up a life that was, it seemed, perfectly designed—a life that moved in a completely different direction from the one where he dropped the ball. We authors, deeming the dropped ball a "tragic event" and "worthy of changing," made a huge ethical mistake in sending Freddy back. Our judgment of the dropped ball as bad served as the catalyst to a changed past, a changed present, and a changed future that even we could not have imagined. Sometimes, you really do need to leave well enough alone.

Freddy could go back and catch the ball, and all would be well if the Multiverse existed, but still, we can assume that in *this* time line, he would still be better off not catching it, and taking the flack from his friends and teammates and those awful parent coaches that have no lives outside of Little League—sorry, got off on a tangent there.

We live in one time line and, until we find a way to branch out, we have to come to grips with the fact that maybe things cannot be fixed by going backward, only by going forward and living a better life with the knowledge and wisdom we bring with us from the past. If we fix the past, we might break the future and leave ourselves with even bigger problems than the one we were trying to fix. And even if we want to go back and change something as trivial as our hair color, or not go to that salon but rather this one here for our manicure, or buy the green car instead of the black one, or even take the red pill instead of the blue pill—we have no idea of how those simple, silly choices might spread out like butterfly wings and cause huge hurricanes later on in our lives, ones we will regret and wish we could go back and fix again.

In his October 2011 interview with *Decoded Science*, physicist Ronald Mallett, whom we profiled earlier in the book, discussed the problem of time travel ethics and the potential misuse of the technology behind it. He referenced going back in time to stop the assassination of President Abraham Lincoln and what race relations might be like if that happened, but he questioned whether or not some kind of regulation would be needed to keep people from going back to the past and causing great harm. "Once the time travel occurs, if someone has altered the past, then everything that we think as being our reality could be an altered reality. So it has to be regulated to make sure we don't do the wrong thing with it." Imagine a new escalation of time travel weapons technology. We haven't even mastered control over the weapons we have that can destroy our future. Could we also destroy our past, and thus our present, if this technology got into the hands of terrorists or dictators bent on power not just over countries and people, but history itself?

Some say the past is dead. But that is not really true. Linear time is an illusion—a figment created by mankind to explain progression and advancement—but to our brains it is a way for us to catalog events and show progress in our lives (as discussed in Chapter 1). We can indeed include the past in that progress, without having to go back and change a thing. Because life is lived in the now, is it not? And no matter what time it is, there really, in the end, is no time like the present.

We wouldn't change a thing. But we still wouldn't mind going back just to take a peek anyway!

As for traveling into the future, how exciting would it be to see how far humanity could reach, and yet what if we saw something we didn't want to see (our own death, the death of a loved one, total apocalypse next Friday at 3 p.m. EST...)? How would we again deal with the knowledge of the future when we come back to the present? Would we change our ways, only to find we also changed the future, but to an even *worse* outcome? Would we, having that knowledge, be able to leave well enough alone and go about our business as if we didn't know what was coming down the pike? Doubtful any of us have that kind of mental and emotional control. Would we try to stop our loved one from dying in that horrible car accident that we saw in our future travels, only to find that, in doing so, we ended up being responsible for the deaths of 10 additional people who might have otherwise lived?

Oh, the ethical questions, the challenges, the what ifs. Yet there is not one of us alive who would pass up the chance to get into a fancy machine and travel back to the past, or forward into the future, because curiosity would get the best of us. Knowing that, we must look to the time when we can do these very things with the understanding that we carry with us, either into the past or the future, a huge responsibility toward our fellow time travelers, toward history, and toward humanity as a whole. We may master the technology and science of time travel long before we master our ability to make good decisions and use our knowledge to the benefit of humanity, and not screw up the planet even more so.

Maybe for a little while longer, we must be content to travel back and forth in time only in fiction, in film and television, and in our dreams and imaginings. But we already travel back in time in another sense when we look through a telescope at objects long dead, seeing only their light just now reaching out from the past and appearing to our human eyes. The cosmic microwave background goes back more than 10 billion years, and that is a long way into the past that technology is allowing us more and more glimpses of with each passing year.

But perhaps the most important reason why time travel is not yet possible has to do more with us, and less with physics and machines: Maybe if we could travel backward and forward in time, we would cease to exist as a species! Bear with us a moment here to explain. As it is, we live most of our lives dwelling on the past and wishing we could change it, or stressing and worrying about the future and what we fear might be around the corner. Perhaps the inability to time travel is another of nature's wonderful survival mechanisms meant to keep us from destroying ourselves. Because here is the deal: If we all want to go back and fix the past, or move forward and see what life has in store for us, no one would ever remain in the present! Simply put, there would be nobody in the here and now to keep things going (well, except for the enlightened few who "get" the present moment, but they would have to breed like crazy to make up for the rest of the missing-in-action!). It's a thought....

Time is so many things. Like the space around us, and like nature, we as humans long to control time because we feel better and more secure when we control our environment. We long to relive great and happy moments from our past and peek into the possibilities of our future. We want to go into the past and watch history unfold, like some Disneyland ride where we simply observe and "ooh" and "ahhh." We want to ride by on a tram and glimpse a panorama view of the future projected on a giant screen before our eyes. We want to go back and fix what we broke, and see what we can avoid breaking in the days to come. We want to master time the way we master the little things in our lives. We want to own time, and not let it own us. Plus we just plain wish we had more time. Maybe that is the rock-bottom truth behind our quest to control, manipulate, and even travel through the landscape of time: When all is said and done about time, maybe the truth is we really just want more of it.

If we ask ourselves why, though, the answer will always come back to this: We all get the same amount of time, at least in this branch of the Multiverse. It's not how much we get that matters, though. It's what we do with it.

Happy trails.

I believe that someday mankind will be able to answer the question, 'What happens when we go back in time and change the past?' Time travel could allow us unprecedented control of our destiny. Ultimately, however, the only thing any of us really have is the present moment.

—Ronald Mallett, *Time Traveler: A Scientist's Personal Mission to Make Time Travel a Reality*

BIBLIOGRAPHY

Bem, Daryl J. "Feeling the Future: Experimental Evidence for Anomalous Retroactive Influences on Cognition and Affect." *Journal of Personality and Social Psychology,* American Psychological Association, 2010.

Boyd, Robert W., and Zhimin Shi. "Optical Physics: How to hide in time." *Nature*, volume 481, November 2011.

Brumfiel, Geoff. "Particles Found to Travel Faster than Speed of Light." *ScientificAmerican.com*, September 22, 2011.

Carroll, Sean. *From Eternity to Here: The Quest for the Ultimate Theory of Time*. New York: Plume, 2010.

Choi, Charles Q. "Leading Light; What Would Faster-Than-Light Neutrinos Mean for Physics?" *ScientificAmerican.com*, October 13, 2011.

Davies, Paul. "A Brief History of the Multiverse." *The New York Times*, April 12, 2003 edition.

———. *How to Build a Time Machine*. New York: Penguin Books, 2003.

———. *About Time: Einstein's Unfinished Revolution*. New York: Simon and Schuster, 1996.

Einstein, Albert. *Relativity: The Special and General Theory*. Charleston, S.C.: Forgotten Books, 2010.

Everett, Allen, and Thomas Roman. *Time Travel and Warp Drives: A Scientific Guide to Shortcuts Through Time and Space*. Chicago, Ill.: University of Chicago Press, 2011.

Gott, J. Richard. *Time Travel in Einstein's Universe: The Physical Possibilities of Travel Through Time*. New York: Mariner Books, 2002.

Greene, Brian. *The Fabric of the Cosmos: Space, Time and the Texture of Reality*. New York: Alfred A. Knopf, 2004.

Hawking, Stephen. "How to Build a Time Machine." *Daily Mail*, UK, December 28, 2010.

Kaku, Michio. *Hyperspace: A Scientific Odyssey Through Parallel Universes, Time Warps and the 10th Dimension*. New York: Anchor Books, 1995.

———. *Physics of the Impossible: A Scientific Exploration into the World of Phasers, Force Fields, Teleportation and Time Travel*. New York: Doubleday, 2008.

Lloyd, Seth, et al. "Closed Timelike Curves via Postselection: Theory and Experimental Test of Consistency." *Physical Review Letters*, volume 106, 2011.

Mallett, Ronald L., and Bruce Henderson. *Time Traveler: A Scientist's Personal Mission to Make Time Travel a Reality*. New York: Basic Books, 2007.

Marrs, Jim. *Above Top Secret: Uncover the Mysteries of the Digital Age*. New York: The Disinformation Company, 2008.

McTaggart, Lynne. *The Field: The Quest for the Secret Force of the Universe*. New York: HarperCollins, 2002.

Moskowitz, Clara. "Warped Physics: 10 Effects of Faster-Than-Light Discovery." *Livescience.com*, September 24, 2011.

Nahin, Paul J. *Time Machines: Time Travel in Physics, Metaphysics, and Science Fiction*. New York: AIP Press, 1999.

———. *Time Travel: A Writer's Guide to the Real Science of Plausible Time Travel*. Baltimore, Md.: Johns Hopkins University Press, 2011.

Palmer, Jason. "Neutrino Experiment Repeat at CERN Finds Same Result." BBC News Online, November 18, 2011.

PBS *Nova.org*. "Sagan on Time Travel." October 28, 2010.

Pickover, Clifford A. *Time: A Traveler's Guide*. New York: Oxford University Press, 2002.

Randles, Jenny. *Breaking the Time Barrier: The Race to Build the First Time Machine*. New York: Paraview Pocket Books, 2005.

———. *Time Storms: Amazing Evidence for Time Warps, Space Rifts and Time Travel*. New York: Berkley Books, 2001.

Susskind, Leonard. *The Cosmic Landscape: String Theory and the Illusion of Intelligent Design*. New York: Little, Brown and Company, 2006.

Tegmark, Max. *Universe or Multiverse?* Cambridge, Mass.: Cambridge University Press, 2007.

Thorne, Kip S. *Black Holes and Time Warps: Einstein's Outrageous Legacy*. New York: W.W. Norton and Company, 1995.

Toomey, David. *The New Time Travelers: A Journey to the Frontiers of Physics*. New York: W.W. Norton and Co., 2007.

Turtledove, Harry, and Martin Greenberg, editors. *The Best Time Travel Stories of the 20th Century: Stories by Arthur C. Clarke, Jack Finney, Joe Haldeman, Ursula K. Le Guin*. New York: Del Rey Books, 2004.

Wilson, Johansson G., et al. "Observation of the Dynamical Casimir Effect in a Superconducting Circuit." *Nature*, volume 479, November 17, 2011.

Wolf, Fred Alan. *Time Loops and Space Twists: How God Created the Universe*. Newburyport, Mass.: Red Wheel/Weiser, 2011.

INDEX

ABOUT THE AUTHORS

Marie D. Jones

Marie D. Jones is the best-selling author of *Destiny Vs. Choice: The Scientific and Spiritual Evidence Behind Fate and Free Will*, *2013: End of Days or a New Beginning—Envisioning the World After the Events of 2012*, *PSIence: How New Discoveries in Quantum Physics and New Science May Explain the Existence of Paranormal Phenomena*, and *Looking for God In All the Wrong Places*. Marie co-authored with her father, geophysicist Dr. John Savino, *Supervolcano: The Catastrophic Event That Changed the Course of Human History*. She is also the co-author of *11:11—The Time Prompt Phenomenon: The Meaning Behind Mysterious Signs, Sequences and Synchronicities*, *The Resonance Key: Exploring the Links Between Vibration, Consciousness and the Zero Point Grid*, *The Déjà vu Enigma: A Journey Through the Anomalies of Mind, Memory and Time*, and *The Trinity Secret: The Power of Three and the Code of Creation* with Larry Flaxman, her partner in ParaExplorers.com, an organization devoted to exploring unknown mysteries. Marie and Larry have also launched the ParaExplorer Series of e-books and articles introducing readers to a variety of subjects. Marie and Larry also host their own radio show, *ParaFringe Radio*, on the LiveParanormal Network.

She has an extensive background in metaphysics, cutting-edge science, and the paranormal and worked as a field investigator for MUFON (Mutual UFO Network) in Los Angeles and San Diego in the 1980s and 1990s. She currently serves as a consultant and director of special projects for ARPAST, the Arkansas Paranormal and Anomalous Studies Team, where she works with ARPAST President Larry Flaxman to develop theories that can tested in the field. Their current project, called The Grid, will be launched in 2012. Marie is a former licensed New Thought/Metaphysics minister and has trained extensively in the Science of Mind/New Thought arena.

Marie has been on television, most recently on the History Channel's *Nostradamus Effect* series, and served as a special UFO/abduction consultant for the 2009 Universal Pictures science fiction movie, *The Fourth Kind*. She has been interviewed on hundreds of radio talk shows all over the world, including *Coast to Coast AM*, *NPR*, *KPBS Radio*, *Dreamland*, *The X-Zone*, *Kevin Smith Show*, *Paranormal Podcast*, *Cut to the Chase*, *Feet 2 the Fire*, *World of the Unexplained*, and *The Shirley MacLaine Show*, and has been featured in dozens of newspapers, magazines, and online publications all over the world. She is a staff writer for *Intrepid Magazine* and a regular contributor to *New Dawn Magazine*, and her essays and articles have appeared in *TAPS ParaMagazine*, *Phenomena*, *Whole Life Times*, *Light Connection*, *Vision*, Beyond Reality, and several popular anthologies, such as *If Women Ruled the World* and five *Chicken Soup for the Soul* books. She has also contributed and co-authored more than 50 inspirational books for New Seasons/PIL.

She has lectured widely at major metaphysical, paranormal, new science, and self-empowerment events, including "Through the Veil," "Queen Mary Weekends," "TAPS Academy Training," "CPAK," and "Paradigm Conference," "Conscious Expo," and "Darkness Radio Events," and is a popular public speaker on the subjects of cutting-edge science, the paranormal, metaphysics, Noetics, and human potential. She speaks often at local metaphysical centers, churches, local libraries, bookstore signings, film festivals and regional meet-ups on writing, the paranormal, human consciousness, science, and metaphysical subjects.

She is also the screenwriter and co-producer of *19 Hz*, a paranormal thriller in development with Bruce Lucas Films, as well as a science fiction feature film titled *Twilight Child*, and she serves as a co-host on the popular Dreamland Radio Show. In her spare time, she raises her son, walks and runs marathons, and is active as a disaster response and preparedness second responder for CERT, the Community Emergency Response Team through CitizenCorps. She is also a licensed ham radio operator (KI6YES).

Larry Flaxman

Larry Flaxman is the best-selling author of *11:11—The Time Prompt Phenomenon: The Meaning Behind Mysterious Signs, Sequences and Synchronicities, The Resonance Key: Exploring the Links Between Vibration, Consciousness and the Zero Point Grid, The Déjà vu Enigma: A Journey Through the Anomalies of Mind, Memory and Time,* and *The Trinity Secret: The Power of Three and the Code of Creation,* with Marie D. Jones, his partner in ParaExplorers.com.

Larry has been actively involved in paranormal research and hands-on field investigation for more than 13 years, and melds his technical, scientific, and investigative backgrounds together for no-nonsense, scientifically objective explanations regarding a variety of anomalous phenomena. He is the president and senior researcher of ARPAST, the Arkansas Paranormal and Anomalous Studies Team, which he founded in February 2007. Under his leadership, ARPAST has become one of the nation's largest and most active paranormal research organizations, with more than 150 members worldwide. Widely respected for his expertise on the proper use of equipment and techniques for conducting a solid investigation, Larry also serves as technical advisor to several paranormal research groups throughout the country.

Larry has appeared on Discovery Channel's *Ghost Lab*, and has been interviewed for dozens of print and online publications, including *The Anomalist, Times Herald News, Jacksonville Patriot, ParaWeb, Current Affairs Herald, Unexplained Magazine, The Petit Jean County Headlight, The Villager Online,* and *The Pine Bluff Commercial*. He has appeared on hundreds of radio programs all over the world, including *Coast to Coast with George Noory, TAPS Family Radio, Encounters Radio, Higher Dimensions, X-Zone, Ghostly Talk, Eerie Radio, Crossroads Paranormal, Binall of America, World of the Unexplained,* and *Haunted Voices*.

Larry is a staff writer for *Intrepid Magazine*, and his work has appeared regularly in *TAPS ParaMagazine, New Dawn Magazine,* and *Phenomena*. He is also a screenwriter with a paranormal thriller, *19 Hz*, in development with Bruce Lucas Films, and a popular public speaker, lecturing widely at paranormal and metaphysical conferences and events all over the country, including major appearances at "Through the Veil," "History, Haunts and Legends," "ESP Weekend at the Crescent Hotel," "The Texas GhostShow," and "DragonCon."

He also speaks widely at local and regional meet-ups, bookstore sign-ings, libraries, and events on the subjects of science, the paranormal, metaphysics, Noetics, and human potential. Larry is also active in the development of cutting-edge, custom-designed equipment for use in the field investigating environmental effects and anomalies that may contribute to our understanding of the paranormal.